Philip Hale

Confessing the Scriptural Christ against Modern Idolatry:

Inspiration, Inerrancy, and Truth in Scientific and Biblical Conflict

©2016 Philip Hale. All Rights Reserved

© 2016 Philip Hale

PUBLISHED BY MERCINATOR PRESS

MERCINATORPRESS@GMAIL.COM

14205 Ida St.
Omaha NE 68142

Scripture quotations are from the ESV© Bible (The Holy Bible, English Standard Version©), copyright 2001 by Crossway, a publishing ministry of Good News Publishers. Used by permission. All rights reserved.

Cover design by Weslie Odom

Hale, Philip.
    Confessing the scriptural Christ against modern idolatry:
      inspiration, inerrancy, and truth in scientific
      and biblical conflict
   Mercinator Press
      Includes bibliographical references and indexes.
      Library of Congress Control Number: 2016906372
      ISBN: 978-0-9975197-0-9

Contents

	Preface	vii
I	A Fundamentalistic Orthodoxy	1
	1 Fundamentalism	3
	2 A Bigger Debate	9
	3 Inerrancy	15
II	The Source of Knowledge	19
	4 Different Truths	21
	5 Denial of Authority	29
	6 Scientific Theology	35
III	A Reverse Historical Comparison	39
	7 The Medieval Doctrine of Scripture	41

8	The Spirit is the Primary Author	49

IV THE SCIENTIFIC SUBJUGATION OF HISTORY   55

9	A New Methodology for Knowledge	57
10	The Sordid History of Historical Consciousness	65
11	Scientific "Facts"	77
12	Methodological Atheism	83
13	Modern A-Theology	91
14	Basis for the Word of God	95

V MODERN ANTHROPOLOGY: REVERSE ARIANISM   101

15	Assumed Incompatibility of Man and God	103
16	Super-Pelagianism	115
17	The Naïveté of Historical Critics	121

VI JUDGING OR BEING JUDGED   127

18	The Sin of Hermeneutics	129
19	Sacred Hermeneutics	135
20	Unity of Attitude toward Scripture and Christ	141
21	Textual Criticism	145

22	The Canon	151
23	Divine Doctrine is the Correct Presupposition	159
24	Proof-Texting: The Only Theological Sin	165

VII	CONSEQUENCES FOR THEOLOGY	177
25	Replacements for *Sola Scriptura*	179
26	Theological Chaos	195

VIII	CHRIST'S NORM FOR THEOLOGY	199
27	Christ and Scripture	201
28	Weak Analogies Prove Nothing	213
29	Historical Criticism in the LCMS: Examples	225
30	Conservative Atheistic Approaches	243

IX	A FOUNDATION FOR CONFESSING	259
31	Timeless Truth	261
32	Meaningless Confessional Subscriptions	269
33	Male/Female Roles Are Divine	275
34	More Christian than Christ?	285
35	Conclusion	291

Acknowledgments 299

APPENDICES 301

Index of Subjects 303

Index of People 309

Index of Bible Verses 312

References 315

Preface

Scripture, its inspiration, and the basis of knowledge have been dominant themes in theology for nearly 400 years. In surveying the field of erature on the inspiration and inerrancy of Scripture, one quickly sees that much has been written that is contradictory, sloppy, and even harmful. Many today would rather avoid the whole issue and not get stuck in this intellectual quagmire, but proceed to "real theology." However, at stake with the doctrine of Scripture is precisely how theology is done and what authority theological statements—including those in sermons, dogmatic texts, and confessional writings—possess. At stake is nothing less than truth itself and the possibility of communicating the truth of Christ to mankind. Therefore, this is not an issue that can be avoided. The Christian must not hide behind scholarly pronouncements or assume that man's opinion will do. Anything less than Christ's Word is insufficient.

The prestige of academic theology, the education of pastors at seminaries based on the university model, and the critical methods of scholarship have much to do with the current confusion over how to define the nature of Scripture. This confusion has not arisen from the Bible itself, but from notions and standards associated with modern science. This particular work endeavors to take a medieval view of theology in contrast to so-called "modern theology"—of which few theologians are presently critical. Undertaking a historical critical inquiry into historical criticism and critical theology itself is much more fruitful than its application to God's words. While this book's confessional stance is staunchly Lutheran, and its main foils are within the Lutheran Church–Missouri Synod (LCMS), the modernist themes of conflict are found in every church and among adherents of every Christian confession.

The journey starts with Scripture, but quickly delves into the intellectual climate that has been in the making for centuries. However, I do not pretend that this analysis is complete, scientific, or scholarly—as if such a thing were possible or godly. The Christian is to confess and bring clarity in light of God's own speech and truth, not hide behind the academic respectability of vague nothings. The doctrine of the inspiration of Scripture is quite simple—it is sufficiently expressed in the Nicene Creed—but this modern culture is so deeply steeped in atheistic assumptions, it is almost impossible to do theology without addressing the intellectual habits of the modern mind and the detrimental effects of modern philosophical currents and Enlightenment assumptions.

It has been said that all theology today is *prolegomena*—that which is said before tackling the subject. While it may not be theology proper, it sets the stage for it, and, if misdone, dooms the theological task before it begins. The definition of theology, its setting in the academic sphere, and the scope of its claims upon mankind are vital questions today, all of which revolve around Scripture and its origin.

My aim here is to show the dangerous effects of false intellectual starting points, that is, how bad *prolegomena* preclude a solid, truthful confession of Christ that can serve as the basis of faith and church. The method of interpreting Scripture and the assumed doctrine of inspiration (or "unspiration," in many cases) radically changes not only a few teachings but the doctrine of Christ and His proclaimed Gospel. That is the thesis of this work, which originally began as a small essay on inspiration, but grew to take on modern theology, scientific-oriented accounts of truth, the general spirit of academic theology, and the false "enlightened" view of man central to modern Western civilization. The Word of God must shine and be scattered among all, even its enemies, especially when it means battling the falsehood that subverts the pure teaching of Christ.

> *Be gracious to me, O God, for man tramples on me;*
> *all day long an attacker oppresses me;*
> *my enemies trample on me all day long,*
> *for many attack me proudly.*
> *When I am afraid,*
> *I put my trust in you.*
> *In God, whose word I praise,*
> *in God I trust; I shall not be afraid.*
> *What can flesh do to me?*
> Ps. 56:1–4

Part I

A Fundamentalistic Orthodoxy

Chapter 1

Fundamentalism

The term "fundamentalism" is prevalent in contemporary religious discourse. Its many suggestive uses illustrate the divergence among those who uphold religion in supposedly similar ways. This is a massive debate. And it is about much more than Scripture's teaching about itself. The modernist–fundamentalist storm centers on authority. Behind the simplistic labels are differing conceptions of God, revelation, man's nature, and the possibility of confessing a timeless truth.

The Nicene Creed is sufficient in its explanation of the fact of biblical inspiration. It reads "[the Holy Spirit] who spoke by the prophets," echoing Lk. 1:70.[1] We cannot say much more than that, because Christ Himself in His inscripturated Word does not. However, the modern morass of books on Scripture and revelation testify that any change to this doctrine is far-reaching, even to the nature of the Gospel itself.

The various modernist attacks on so-called "fundamentalisms" show that the true nature of the debate is bigger than Scripture. Fundamentalism is said to encompass more religions than just Christianity. This illustrates that the debate is more fundamental and deep-seated than the interpretation of the Christian Scriptures or the specific doctrines propagated by any one religion. Islamic "fundamentalists" and Jewish "fundamentalists" do not claim to have the same God as Christians, but are all lumped into one big fundamentalist category.

---

[1]The Nicene Creed in Latin reads: *qui locutus est per prophetas*, while Lk. 1:70 in the Vulgate reads: *locutus est per os sanctorum*.

A Fundamentalistic Orthodoxy

The definitions and characterizations of fundamentalism are legion. To correctly answer the question, "what is fundamentalism?" is to resolve the modern controversy on Scripture and revelation to a large degree. However, many uses of the fundamentalist label are simply pejorative and insulting, meaning little to nothing. But the basic modernist–fundamentalist divide shows the decisive and stark split in theological understandings that has been developing for more than 400 years.

Fundamentalism suffers from the charge of being simultaneously anti-intellectual yet rationalistic and philosophical. It is said to be traditional, dogmatic, and outdated—yet also, somehow, reactionary, violent, and divisive. The basic charges against fundamentalism are contradictory. "The traditional fundamentalist understanding of the [Bible] derives from reason and philosophy rather than from sympathy with biblical insight."[2] The chasm between the two sides is so great, and the basic assumptions so ingrained, that the real differences are almost unintelligible to those on either side. The modernist and fundamentalist factions think in fundamentally incompatible ways. The view of anthropology, nature, communication, and truth are the real issues, and they are, of course, intimately bound up with Christ Himself, the Creator and Redeemer.

"Fundamentalism" is "a theological swearword" used by modernists to condemn religious orthodoxies.[3] By asserting that fundamentalist convictions are illogical and intellectually segregating the anti-modernists, the modernists are claiming to be right or "orthodox." Fundamentalism has a hard, factual definition. As an early 20th-century American movement, fundamentalism is defined by a narrowing of important doctrines to a few specific fundamentals. The reduction in doctrinal content is not the offense to modernism. The fact that fundamentalists believe anything, despite whatever divisions or consequences it may bring, is the pill modernism cannot swallow.

Modernism uses a different method to obtain truth and to prove its deeply held convictions. Modernist, or Enlightenment-induced, thinking is self-supported by technological improvements. It can claim the observable results of toleration based on the principles of natural reason.

---

[2] James Barr, *Beyond Fundamentalism: Biblical Foundations for Evangelical Christianity* (Philadelphia: Westminster Press, 1984), 37.

[3] Clark H. Pinnock in *The Fundamentalist Phenomenon: A View from Within; A Response from Without*, ed. Norman J. Cohen (Grand Rapids: Eerdmans, 1990), 40.

The evidence seems to indicate to modernists that their orthodoxy is the correct one, since they allow only empirical proof. Fundamentalists, however, operate with an incompatible paradigm for interpreting reality and verifying truth claims. To them, persecution and minority-status are evidence that they are orthodox.

Consider the audacious charges made against the most thorough theologian of the Lutheran Church–Missouri Synod (LCMS), whose three volume *Christian Dogmatics* is still the standard text more than 90 years after it was written. An article by Leigh D. Jordahl appeared in 1971 titled: "The Theology of Franz Pieper: A Resource for Fundamentalistic Thought Modes among American Lutherans." Pieper, according to Jordahl, held to a doctrine of inspiration "inherited from the Princeton theology" of the mid-to-late 19th century.[4] But Princeton theology was confessionally Reformed, not Lutheran. Many of its teachings are condemned explicitly by Pieper and all confessional Lutherans holding to the *Formula of Concord*.

This type of evolutionary blunder is endemic among modernists who cannot fathom a great intellect submitting fully to a supernatural revelation. A driving modernist urge is to pinpoint an idea to a specific historical origin, which immediately disqualifies it as man-made. Fundamentalism is said to be "a mood of intransigence legitimized and given timeless validity by a dogma of change as heresy," which is "identified with a specific hermeneutic, and employed in order to insulate oneself or one's group from the forces of modernity."[5] It is defined not as having any valid substance itself, but as a mindless reaction against progress and true knowledge.

"Had Pieper's experiences been different he might have become a social gospeler."[6] The arrogance of this statement is astounding. Despite Pieper's own claims to appeal to divine truth, Jordahl relativizes his entire theology, because it is based on a presumably faulty experience. The root issue is not Scripture or any theological conclusion. Instead, outside influence, in the name of scientific knowledge, is held as the

---

[4]Leigh D. Jordahl, "The Theology of Franz Pieper: A Resource for Fundamentalistic Thought Modes among American Lutherans," *Lutheran Quarterly* 23 (1971), 118–119.

[5]L. Jordahl, "The Theology of Franz Pieper," 122.

[6]L. Jordahl, "The Theology of Franz Pieper," 125.

ultimate and final truth for modernists.

To modern man, the appeal to divinely revealed truth is laughable and simply the product of a petty, deranged mind of limited experience. One Roman Catholic scholar attacked some of those belonging to his own church body, claiming that "Catholic fundamentalism is a psychosocial disease." The evidence of this truth is established, in his mind, by an absurd intransigence of "the 'right-to-life' movement in reaction to the 1972 Supreme Court decision legalizing abortion."[7] An obsession with the commandment "you shall not murder," is unreasonable, evidently, because certain interpretations of this command can cause great emotional distress and societal unrest.

What specific doctrine does modernism strike at most violently? "In the 17th century this doctrine [of inspiration] could be affirmed with a minimum of scientific embarrassment. A lot has changed in two centuries."[8] More than 350 pages in Pieper's dogmatic text on "the nature and character of theology" and "Holy Scripture" are dismissed with a modernist sleight of hand as irrelevant. Another scholar speaks of "the verbal inspiration of the Lutheran Fundamentalists."[9] Fundamentalism and modernism are vague categories, but to the extent they are revealing, they operate with different conceptions of authority and methods of discovering truths.

Pieper's great defense of Scripture is precisely against all modernist tendencies. He took the threat of modern theology very seriously. Modernism bases certainty and knowledge on rational methodology. Its principles unavoidably conflict with Scripture. In contrast, fundamentalism is a "simple and child-like acceptance of Scripture."[10] Christian fundamentalists are those who hold that the Bible is "true in all particulars for moderns as well as ancients."[11] Both sides seem to agree that the doctrine of Scripture, specifically the reach of its claims, are at the

---

[7] Patrick M. Arnold in *The Fundamentalist Phenomenon,* ed. N. Cohen, 179.

[8] L. Jordahl, "The Theology of Franz Pieper," 127.

[9] Miikka Ruokanen, *Doctrina divinitus inspirata: Martin Luther's Position in the Ecumenical Problem of Biblical Inspiration* (Helsinki: Luther-Agricola Society, 1985), 115.

[10] G. C. Berkouwer, quoted in: *Inerrancy and the Church,* ed. John D. Hannah (Chicago: Moody Press, 1984), 397.

[11] Stephen J. Nichols and Eric T. Brandt, *Ancient Word, Changing Worlds: The Doctrine of Scripture in a Modern Age* (Wheaton, IL: Crossway, 2009), 26.

forefront of the intellectual divide. The two camps have different underlying conceptions of authority. All religions claiming absolute authority, including Christianity, are caught up in the prevailing cultural forces of modernism.

Chapter 2

A Bigger Debate

Simply speaking about Scripture's use and nature is not satisfying. It is not theology proper; however, false academic approaches call for spelling out an explicitly Christian approach to the Bible. The contents of Scripture will be misunderstood if it is read in a godless way.

As a result of cultural and intellectual patterns, theology has gravitated toward being almost exclusively concerned with pre-theological matters. In the old orthodox dogmatics, this topic of the nature of theological claims and the use of Scripture in relation to theology was called *prolegomena*, that is, "the things spoken beforehand." But because Satan has most virulently attacked the Church in the area of *prolegomena*, Christians can by no means bypass this debate. To avoid the issue is to be unwittingly caught up in destructive heresies that are no less harmful than the Trinitarian and Christological controversies faced by the Early Church. In fact, the modernist heresy is at least as far-reaching as the errors described in the Ecumenical Creeds. It casts doubt on the very possibility of knowing truth and articulating it. Therefore, meaningful theological work cannot avoid the inerrancy debate.

The doctrine of inerrancy is given various erroneous origins. Besides the association with 19th-century Reformed theologians, many scholars hold that "the belief in inerrancy is novel, that it is a product of post-Reformation Protestant Scholasticism."[1] An errorless biblical text seems

---

[1] *Inerrancy and the Church,* ed. J. D. Hannah, viii.

to be a labeled a Protestant error, perhaps because this branch of Christendom is not explicitly tied to tradition as a theological source. Their supposedly new Reformation doctrine of Scripture alone [*sola Scriptura*] makes them easy targets.

The facts of history demonstrate that positing inerrancy as a property of Scripture is not new, nor exclusively a Protestant claim. In 1893 Pope Leo XIII stated in the encyclical "*Providentissimus Deus*: On the Doctrine of the Modernists":

> For all the books which the Church receives as sacred and canonical, are written wholly and entirely, with all their parts, at the dictation of the Holy Ghost; and so far is it from being possible that any error can co-exist with inspiration, that inspiration not only is essentially incompatible with error, but excludes and rejects it as absolutely and necessarily as it is impossible that God Himself, the supreme Truth, can utter that which is not true.

No defense of the Bible's inerrancy and inspiration could be stronger than this statement by an opponent of *sola Scriptura*.

Moreover, it is claimed that, following the pattern of the Nicene Creed, "God, Who spoke first by the Prophets, then by His own mouth, and lastly by the Apostles, composed also the Canonical Scriptures, and that these are His own oracles and words . . . that God Himself has composed." Modernists must conclude that this statement by a Roman pope is fundamentalistic. While the time frame of this defense of Scripture's inerrancy is within that of Princeton's B. B. Warfield (d. 1921), the notable Protestant defender of Scripture, there is no provable connection between this Reformed theologian and the Roman pontiff. Perhaps the rise of fundamentalism is not new at all, but a proportional response to a universal error attacking all confessions. A later Roman pope indicates exactly this: "no one will be surprised that We should define [modernism] to be the synthesis of all heresies."[2]

No matter how distasteful the implications of the doctrine of Scripture, this controversy is unavoidable. The first president of the LCMS, C. F. W. Walther, said in his 1885–1886 evening lectures on inspiration, "this doctrine of inspiration belongs to the burning theological questions today and among these is without doubt the most significant," because

---

[2] Pope Pius X, "*Pascendi Dominici Gregis:* On the Doctrine of the Modernists," encyclical, 1907 (http://www.papalencyclicals.net/Pius10/p10pasce.htm).

it is "the basis of all other doctrines, on which they are raised up."³ Yet, at the same seminary at which he gave those lectures, another professor later said: "inerrancy makes an idol of Scripture."⁴ Though a fairly minor point in itself, inerrancy has become the fundamental watershed in theology. The controversy has not subsided in the intervening years. Well over a century old, the debate over Scripture rages just as intensely. The modernist–fundamentalist divide reveals that there are two gods vying for the heart of man.

Inspiration and inerrancy are at the forefront of the modernist attack. Reducing the positive content of these particular words is critical to the well-being of modernist thought. "We used to think of inspiration as a procedure which produced a book guaranteed in all its parts against error and containing from beginning to end a unanimous system of truth. No well-instructed mind can hold that now."⁵ The issue is even bigger than the source of theology—thus it encompasses all orthodoxies, even those outside of Christianity. Whether religious doctrine is from God or of man is foundational to defining the truth of theology, and clarifying the modernist–fundamentalist debate.

The attack on the previously universal assertion of Scripture's inerrancy demonstrates an entirely new theological approach. Modernists question or deny that God can communicate. To them inspiration implies "a yoke of bondage to an exploded relic of post-Reformation scholasticism."⁶ To the modern thinker, religious truth must be erring because truth has been redefined. Here is the root of the hatred of the Spirit's unerring inspiration of Scripture: if God does not speak clearly, man speaks for Him in a more palatable way.

One modernist defines fundamentalism as "the assumption that what the Bible says on controverted ethical questions is clear (it usually is not) or valid for all time."⁷ This quote actually misses the point. It is

---

³C. F. W. Walther, "Walther's Evening Lectures on Inspiration, 1885–1886," trans. Thomas Manteufel (Presented to the Walther Round Table, 2005–2007; http://www.lutheranhistory.org/waltherrt/wrt-inspiration.htm), Lecture I.

⁴Paul A. Zimmerman, *A Seminary in Crisis: The Inside Story of the Preus Fact Finding Committee* (St. Louis: Concordia Publishing House, 2007), 176.

⁵Harry Emerson Fosdick, quoted in: S. J. Nichols and E. T. Brandt, *Ancient Word, Changing Worlds*, 54.

⁶Henry Preserved Smith, *Inspiration and Inerrancy: A History and a Defense* (Cincinnati: Robert Clarke & Co, 1893), 87.

⁷"The [liberal] ELCA [Evangelical Lutheran Church in America] is still trying

not only the *controverted* issues that have supra-human authority. Every issue to which Scripture speaks has timeless currency. Fundamentalists of all stripes hold to an unlimited, unchanging truth above every person and culture.

The modernist side resists timeless authority by claiming that what God says is conditioned on man: "the Lord does not through the prophets utter perfect, final, or ultimate and unchangeable statements."[8] The difference between modernism and fundamentalism is best described by their different bases for certainty and, therefore, ultimate grounds for truth claims. "Philosophy grounds itself on a commitment to epistemological [man-based] certainty, while revelation is rooted in a classic revelation confirmed by tradition, prophecies, and miracles. One is open, critical, and relative; the other is closed, obedient, and passive. In the end, there is no way the two can be reconciled, no matter how vigorous the attempt."[9] The fundamentalist approach to knowledge is based on supernatural knowledge which is above, and likely contrary to, rational understanding. Modernism, on the other hand, is man-centered in its perspective, and views religion (and truth) in terms of adaptation and synthesis. "The religion that ceases to change ceases to remain the same."[10] It conceives of certain knowledge as based on a naturalistic system. Its basis and standard of truth move along with the changing culture. One is religious in scope involving submission, the other is anti-religious depending on reason's critical apprehension of facts. A 1924 *Christian Century* article made this claim: "The God of the fundamentalist is one god; the God of the modernist is another."[11]

The debate is often styled as ignorant or simple-minded Protestants versus those who think and do intellectually meaningful theology. But the following confession by a thoughtful Eastern Orthodox monk shows otherwise: "Scientific conceptions lie within the sphere of relative knowl-

---

to break out of its own version of fundamentalism." Ralph Klein, "How My Mind Has Changed," Conference: Building the One Foundation: Seminex at 35 (2009; http://www.lstc.edu/media/seminex/06-24-09-faculty-panel-Klein.pdf), 3.

[8] J. Barr, *Beyond Fundamentalism*, 24.

[9] Michael L. Morgan, *Dilemmas in Modern Jewish Thought: The Dialectics of Revelation and History* (Bloomington: Indiana University Press, 1992), 66.

[10] R. Klein, "How My Mind Has Changed," 3.

[11] "Fundamentalism and Modernism: Two Religions," quoted in: Chester E. Tulga, *The Case against Modernism* (Chicago: Conservative Baptist Fellowship, 1949), 57.

edge and are always subject to uncertainty and change, whereas the dogmatic, theological teaching of the Church rests on the certainty of Divine revelation and does not change." The issue is much bigger than Protestantism or the definition of inerrancy. "God is the Author of all truth, and anything genuinely true in Scripture cannot contradict anything that is genuinely true in science."[12]

These two approaches to knowledge and truth clash most strongly in Christendom over the definition, meaning, and import of the doctrine of inspiration. Modernists hold that Scripture can be held as authoritative without inerrancy. But inerrancy, if appropriately redefined, is not even a hindrance. Fundamentalists, though, conceive of truth as stable and unchanging. They have a strict definition of inspiration and inerrancy and see them as vital teachings.

Modernists and fundamentalists show the incompatibility of their assumptions most completely in the origin of Scripture. Their different ways of perceiving and verifying knowledge diverge in the supernatural claims of divine revelation. Fundamentalists rely on revealed knowledge, not immediately available to all. Modernists generally rely on scientific methods to verify knowledge, making divine communications improbable at best. The word "science" often closely follows modernist definitions of inspiration. In religion, "often one finds notions of revelation and inspiration, and hence of normative authority, that cannot be easily reconciled with the procedures of science."[13]

Why has inerrancy in God's communication become problematic? God's ability to communicate, to reveal anything, is thought absurd. Since inerrancy—that God can speak correctly and without limitation in human language—is part and parcel of Scripture's inspiration, it has borne the brunt of the modernist attack. Inspiration is easy to redefine in virtually meaningless ways, but inerrancy is a more solid word in our vocabulary, therefore it endures the most vitriolic attacks. This only illustrates that the two sides have radically incompatible conceptions of the divine.

---

[12]Seraphim Rose, *Genesis, Creation, and Early Man: The Orthodox Christian Vision,* ed. H. Damascene (Platina, CA: St. Herman of Alaska Brotherhood; rev. ed., 2011), 77, 116.

[13]Holmes Rolston III, *Science and Religion: A Critical Survey* (Philadelphia: Temple University Press, 1987), 6.

# Chapter 3

Inerrancy

A confession of the inerrancy of Scripture is a very limited statement. It is in no way a complete view of Scripture or indicative of Christian faith. Although correct, it asserts only that the Bible is true—saying more about one's view of truth than about the particular content of Scripture.

Confessing Scripture's inerrancy still falls short of asserting that it is God's actual Word. Inerrancy (correctness) does not imply a divine communication. A message that is originally erring, though, surely rules out the possibility that it is of divine origin. The inerrancy of the actual writing claiming to be revelation has become the dividing line. Human words can be inerrant, but divine words cannot be errant. The issue is not correctness in general, but the specific areas in which the two ways of determining truth conflict. It is in essence a battle between two truths, and their ultimate grounds.

The truth of Scripture cannot be in error to the Christian who accepts it as pure revelation. This is an *a priori*, an assumed prejudgment, apart from man's verification or analysis. Inerrant doctrines are absolute presuppositions, which rule and dictate beforehand what is, and is not, acceptable. It means no facts can be established by man against these most solid truths. To even question a revealed truth is disobedience against the Revealer.

Inerrancy is not a specifically Christian doctrine, it is true, but that property is necessary for Scripture to be confessed as God's book.

A Fundamentalistic Orthodoxy

The "notion of inerrancy . . . rests ultimately on truth and falsity."[1] Inerrancy is like shingles on the roof of the Church. While far from the foundation of Christ Jesus, the whole structure will rot and succumb to decay eventually, if the Scriptures are no longer perceived to rule man inerrantly. One's view of Scripture is at the same time a definition of theological truth and the basis for securing all the doctrines it contains.

Augustine's (d. 354) ancient dictum is at odds with modernism: "I have learned to pay [the Scriptures] such honor and respect as to believe most firmly that not one of the authors has erred in writing anything at all."[2] Was one of the greatest theologians of the Early Church a fundamentalist? The fundamentalist designation, historically speaking, does not make sense. That is because modernism, as its name suggests, is the novel one. "It is we [moderns] who have departed from the tradition, not [the fundamentalists]. . . . The Bible and the *corpus theologicum* [body of theology] of the Church is on the fundamentalist side."[3]

The modernist sees any fundamentalism as a psychic deformity, a love of militantism, and a pointless clinging to irrelevancies. But that is because both sides work with different versions of truth, and even language itself. Fundamentalism is "a designation for any and every kind of orthodoxy within all the world religions."[4] How can a traditional orthodoxy be an " 'extremism,' 'fanaticism,' even 'terrorism' "?[5] Only in the face of the most extreme ungodliness, known as modernism. Its basic tenets, rooted in Enlightenment-era philosophy, will not allow any compromise with a fundamentalist truth, that is, a communicating deity. Errant truth and erring divine speech have always been considered incompatible with the God of the Bible. An errant word cannot be the Word of God in any meaningful sense.

But what has made revelation impossible for moderns? Inerrancy, "verbal dictation . . . [and] uniformity of doctrine . . . have become

---

[1] John S. Feinberg in *Inerrancy*, ed. Norman L. Geisler (Grand Rapids: Zondervan, 1980), 166.

[2] Letter 82, quoted in: *Inerrancy and the Church*, ed. J. D. Hannah, 48.

[3] Kirsopp Lake (1926), quoted in: *Challenges to Inerrancy: A Theological Response*, eds. Gordon Russell Lewis and Bruce A. Demarest (Chicago: Moody Press, 1984), 350.

[4] Jaroslav Pelikan in *The Fundamentalist Phenomenon*, ed. N. Cohen, 3.

[5] Riffat Hassan in *The Fundamentalist Phenomenon*, ed. N. Cohen, 157.

incredible in face of the facts."[6] One should rightly ask, from where do facts come from that change the doctrine of Scripture and potentially the whole dogmatic structure of Christianity?

---

[6]Harry Emerson Fosdick, quoted in: Theodore Engelder, "Verbal Inspiration: A Stumbling Block for the Jews and Foolishness to the Greeks," 16 parts, *Concordia Theological Monthly* 12:4–13:12 (April 1941–Dec. 1942), Part 1, 248.

Part II

The Source of Knowledge

Chapter 4

Different Truths

Modernist critics usually admit that the scriptural contents themselves are not their main problem. Rather, it is outside knowledge, assumed to be true, that has altered their view and use of the Bible. The very possibility of divine statements conflicting with the more fundamental truth of nature is the root of the discomfort over Scripture's inerrancy. Physical facts are given priority over what is said to be divine revelation. When "the fundamentalist takes revelation to be identical with the propositions of the biblical text . . . he is in direct contradiction with modern science."[1] Those theologians are at least honest who admit that science is their baseline truth when reading Scripture's claims.

The offense of inerrancy is specifically in "this insistence on the Bible's inerrancy in history and scientific matters."[2] The starting point and final arbiter of truth, for moderns, is that which is established by scientific means. What was common thinking, that God did not lie in His speaking, became problematic in the modern era. Inerrancy, "in a sense seemed to risk the *whole Christian faith* upon one error."[3] The change was in man, not God or His Scripture. Truth became scientific, and, therefore, based on observation. Inerrancy is viewed by moderns as a

---

[1] James Barr, quoted in: *Inerrancy*, ed. N. L. Geisler, 108.
[2] Warren Quanbeck, quoted in: John Warwick Montgomery, *Crisis in Lutheran Theology: The Validity and Relevance of Historic Lutheranism vs. Its Contemporary Rivals*, vol. 1 (Grand Rapids: Baker Book House, 1967), 18.
[3] Ernest R. Sandeen, quoted in: *Challenges to Inerrancy,* eds. G. R. Lewis and B. A. Demarest, 363.

burdensome and impossible claim. Truth claims must be proved certain by the reasoning modern man, not accepted in ignorant submission. Man through new methods became the key ingredient to all truth, replacing the traditional role of the deity.

The great chasm is between naturally apprehended factual knowledge and passively received religious revelation. That true facts cannot be proved by scientific means does not make sense to moderns.

> I'm surprised whenever I encounter a religious scientist.... How can a bench-hazed Ph.D. ... go home, read in a two-thousand-year-old chronicle, riddled with internal contradictions, of a meta-Nobel discovery like "Resurrection from the Dead," and say, gee, that sounds convincing? Doesn't the good doctor wonder what the control group looked like?[4]

The issue is whether science, and its driver, man, is unbounded in the ability to discover truth. Is its approach viable for all truth, or limited to things of nature, as a purely naturalistic method? The very fact that all approaches labeled "scientific" are assumed to be without bias and prejudice is a telling sign. The underlying assumptions of science are rarely examined. Scientific knowledge has completely overwhelmed all other possible means of truth, so that to critique its validity is unthinkable.

While science and religion are opposed in the minds of modernists, the facts do not bear it out. There are many physical scientists who understand the field well as its practitioners, and yet are quite religious. "Scientists hold no religious view uniquely their own." Many are even self-identified fundamentalists. "Wrong are those who hold that a 'scientific spirit' implies denial of all devotion to religious creeds."[5]

One may know science, but see the scientific method as having a narrow and definite sphere of validity. It is not a way to ultimate truth for all scientists. But the perceived conflict between natural truth and revealed truth, among non-scientific minds, is a key feature of modernism. A fundamentalist pope of a century ago asserted: "There can never, indeed, be any real discrepancy between the theologian and

---

[4]Natalie Angier, *My God Problem*, (http://edge.org/conversation/my-god-problem, Nov. 19th, 2006).

[5]Edward LeRoy Long Jr., *Religious Beliefs of American Scientists* (Philadelphia: Westminster Press, 1952; reprint, Westport, CT: Greenwood Press, 1971), 145–146.

the physicist."⁶

The issues involved with inerrancy are wide-sweeping. They are illustrated in this statement: "The claim of inerrancy can only cause a smile on the part of one resolved to retain his own mind."⁷ Biblical texts, and their interpretation, are not the issue. Rather, it is the method of validating truth statements that affects the principles of theology and define its domain. The entire Christian religion and the very definition of truth are at stake in the arguments of fundamentalists and modernists.

Fundamentalism, in itself a modern category, is not intrinsically Christian. The assumption of an erring Bible is just one facet of the modernist attack against all authority, religious and secular. This much larger problem must affect all authoritative doctrine, including the authority of Christ Himself. "The dominant [Christian] fundamentalist understanding of Jesus, which insists on the definition of him as being God, is actually based on a rather thin (or possibly non-existent) line of New Testament evidence and ignores the main line."⁸ The doctrine of Scripture, especially its inspiration, affects how all doctrine is approached. One's position on the inerrancy of Scripture reveals one's actual theological norm and basis. The extent of religious authority determines whether theology is man-made and open to development or divine and unchanging.

Once the starting point of theology is changed, the ending point must also change, in the unfortunate case of consistent thinkers. Most pre-modern heresies attacked a single doctrine directly. This modernist heresy, the synthesis of all heresies, attacks every single one, by undermining the foundation for all Christian thinking and speaking.

The Bible itself can be extolled and used extensively by modernists, but its identification with inerrant, propositional revelation is ruled out. The propositional quality refers to "the truth-content extractable from Holy Scripture," which can be directly applied today without being culturally translated or contextualized.⁹ Biblical inerrancy means

---

⁶Pope Leo XIII, "*Providentissimus Deus*: On the Doctrine of the Modernists," 18.

⁷James Y. Simpson, *Landmarks in the Struggle between Science and Religion* (London: Kennikat Press, 1925; reprint, 1971), 136.

⁸J. Barr, *Beyond Fundamentalism*, 60.

⁹Clark H. Pinnock, *A Defense of Biblical Infallibility* (Phillipsburg, NJ: Presbyterian and Reformed Publishing Company, 1967), 4.

that the claims of Scripture automatically override all scientifically-determined statements of truth. When moderns read a book full of typographical errors, is it discredited? Moderns are overly preoccupied with critiquing and judging the visible evidence from their own vantage point, not God's. A special category is therefore made for "religious truth," so that it is immune from rational criticism, but also, as a side effect, from cognitive belief and rational apprehension. Truth has not changed, but in modern terms real truth, dealing with facts and propositions, is reserved for scientific ventures: "science replaced theology as the standard of knowledge."[10]

The modernist side does not denounce Scripture, though it can only appear that way to the fundamentalist who holds the Bible to be the highest authority—an authority identical with God Himself. Scripture is rather uplifted in modernist thought by extolling "dynamically" the presumably errant "truth" that can still lead powerfully to salvation. Bare authoritative facts lead to dry intellectualism, or dead orthodoxy, it is claimed.

The playing of the dynamic versus the static aspects of revelation is unbiblical and unhistorical. "It was the writers of the period 1700 to 1860 who brought into opposition the propositional and dynamic views of revelation."[11] For example, an LCMS pastor claims: "Simply put, the phrase, 'authority of Scripture' can only make theological sense if it is understood in functional terms." When the static truthfulness of Scripture is denied, it is futile to retreat to the dynamic-only conception of Scripture. "Authority is not so much an ontological property of the biblical writings . . . but it is an *activity* of the Triune God."[12] A presently acting truth that is not propositional cannot be proved wrong or criticized by rational man. It does not conflict with any piece of modern knowledge, because it cannot even be rationally grasped.

While sounding impressive, an activity-only truth cannot be read, described, preached, taught, or confessed. It has nothing to do with rational man. It is irrationalism—a contentless Christianity. Contrary

---

[10] S. Rose, *Genesis, Creation, and Early Man*, 535.

[11] H. D. McDonald, *Theories of Revelation: An Historical Study 1700–1960* (Grand Rapids: Baker Books, 1979), 1:272.

[12] Peter H. Nafzger, *These Are Written: Toward a Cruciform Theology of Scripture* (St. Louis: Wipf and Stock, 2013), 25, 135.

to this sort of Barthianism,[13] one should say, "God gave me my reason and all my senses." The scriptural Gospel is not irrational or a retreat from hard facts, but neither is it generated or measured by man's sinful reason.

As with the Lord's Supper, one can effortlessly praise Scripture and its benefits, but total emphasis on the possible activity of God is an evasion of confessing what it is, in propositional form. "The bread and wine of the Lord's Supper is Christ's body and blood" is a propositional statement. But by using only dynamic terms, which speak of its effects, one may avoid a clear confession of what it actually is: "God works powerfully in the Supper." This uplifts the Supper, but says nothing one can verbally relate to other doctrines, such as faith and sin. The property of inerrancy clarifies how the Bible relates to other accepted statements of truth.

The dynamic aspect of Scripture is not in danger. A fundamentalist sees no opposition between what Scripture is and what it does. In fact, they are intimately connected. That Scripture describes facts of history and nature does not conflict with its promises in Christ. The promises of Christ are based on historical facts, especially His death and resurrection, which can be propositionally stated. The modernist, though, plays static facts and present events against each other, because only the present has validity for him. As with the Lutheran approach to the Supper, one must be able to answer the question: "What does the unbeliever read and hear when presented with the Scriptures?" God speaking can be the only answer of orthodoxy. There are no qualms about any topic found in Scripture, because its author, God, is trustworthy. "The fact that Scripture is not a text-book of science has no bearing on the question whether its scientific statements are true."[14] God chose to give truth in a non-academic manner and non-scientific language.

The intentionalist, or functional, view of truth shifts the focus from words to unknowable intentions. An errant truth is a withdrawal from facts and from the world. This requires a complete shift in how knowledge is acquired and facts are authenticated. Behind seemingly minor statements on the doctrine of Scripture are issues of great concern, down

---

[13]This theological school is named after the immensely influential neo-orthodox Swiss theologian Karl Barth (1886–1968).

[14]T. Engelder, "Verbal Inspiration: A Stumbling Block," Part 7, 897.

to the very nature of truth.

What Scripture does and its effects cannot replace a confession of what it is, since this determines the method of theology. If Scripture or the Gospel is made immune to human criticism by retreating from universal facts, divine truth itself is no longer human. Saying that God can work through error is irrelevant if God Himself tells lies in actuality or does not speak at all. At stake is the character of God—is He errant or inerrant? An unedited, error-riddled science text-book full of typos would not be praised as a vehicle for true statements or knowledge. In Scripture, we have words and statements describing facts, not dynamic intentions or vague, wordless ideas.

The inspiration of Scripture has to do with its authority, from the time of its writing. The Bible's words had a beginning, and this origin, whether ultimately of God or men, establishes its authority. The dynamic power of Scripture is irrelevant to the greater issue of Scripture's authority. God's Word preached by a man also has dynamic power to save, but it is un-Christian to measure Scripture by man's thoughts. No sermon can establish the doctrine of Christ, nor do human words used as God's instruments mean total inerrancy. The judging status of Scripture speaks to its truthfulness and role in establishing knowledge, not merely its effective power or instrumental use.

"Inerrancy" is an abstract word, denoting a general concept which does not immediately lend itself to the proclamation of Christ. But when the dominant heresy undermines all language and truth, a blunt and unequivocal word is needed to confess the truth. To fail to confess inerrancy is to capitulate to the pagan philosophy of modernism. To deny the truth of Scripture is to deny God's truthfulness, or to put God so far away that He cannot have any contact with mankind. Of course, inerrancy is by no means sufficient for confessing the doctrine of Scripture. But there should be no embarrassment over it, if one holds to the inerrant Christ.

True, historical fundamentalism of the early 20th century, with its five doctrinal emphases, did not take Scripture seriously enough. Lutherans do not just stress inerrancy on the most controverted articles of the faith but every teaching of Scripture. Inerrancy just happens to be an essential property of divine authority. It is an essential ingredient to any truth, let alone divine truth. Inerrancy is not specifically Christian, but neither are words, utterances, and an accounting of facts.

Inerrancy speaks to rationality and coherence—to what degree statements are factual and can be related to other statements. To moderns arguments based on revelation sound rationalistic. However, it is not complexity or long meandering arguments that are characteristic of modern, scientific rationalism. Modern rationalism refuses to acknowledge boundaries to man's field of knowledge and reason's inability to comprehend divine matters. "Rationalism . . . is a method rather than a doctrine; an unconscious assumption rather than a principle."[15]

The inerrancy of the Bible is problematic to moderns, because a new, superior inerrancy was "discovered"—the light of reason. "To consider ourselves bound by reason alone, and by no other human court exterior to us, is to believe in our own infallibility, and ourselves alone."[16] This explains the offense of inerrant biblical propositions. When they come from Scripture, they demand that the hearer submit, and if inerrant, they can never be proved wrong by rational means. Propositions remain true outside a person's vantage point or examination. Divine propositions, or doctrines, do not admit criticism, or even questioning, when backed by the full, undiluted authority of God.

The Bible's statements on idols seem outdated. That is because all external gods, true and false, have been made obsolete by modernism. Worship of human reason has replaced physical idols. The rationalist assumes: "Our reason is manifestly God in us."[17] There is even a strong case to be made that this modern rationalism came from within Christianity itself. The true Gospel was perverted and turned into a secular religion.

Martin Luther had a very negative view of reason's role in theology. He taught the scriptural view of not just the weakness of reason but its active rebellion and hatred of the true God. In one of his last sermons he reiterated this by calling reason "the Devil's whore."[18] In spiritual

---

[15] G. R. Cragg, *Reason and Authority in the Eighteenth Century* (Cambridge: Cambridge University Press, 1964), 28.

[16] Frédéric Bettex, *The Bible the Word of God,* trans. from 3rd. German ed. (New York: Eaton and Mains, 1904), 22.

[17] Josias Friedrich Christian Loeffler (d. 1816), quoted in: Theodore Engelder, *Reason or Revelation* (St. Louis: Concordia Publishing House, 1941), 14.

[18] Quoted in: B. A. Gerrish, *Grace and Reason: A Study in the Theology of Luther* (Oxford: Clarendon Press, 1962), 1.

matters, where reason is worse than blind, it refuses the truth. "For the Spirit is required for the understanding of Scripture, both as a whole and in any part of it."[19]

Reason is divinely created and has much value in temporal matters. It has natural authority over the created things of this world (Gen. 1:28). But no conclusions from reason are allowed in spiritual matters.[20]

If God does not reveal propositions, we cannot say anything definite about Him. What replaces the traditional trust in God's words? Man's reason is the assumed infallible authority for modernism. Science is not divine revelation, but a method for gaining knowledge that starts with trust in oneself and distrust of external prejudices. If theology and its divine source are not defended against the rationalist critique, divine knowledge will be built on the necessarily errant scientific mindset and swept away as meaningless and ultimately nonsensical.

---

[19]Luther, *The Bondage of the Will* (1525), *Luther's Works,* eds. Jaroslav Pelikan and Helmut Lehmann, 56 vols. (St. Louis: Concordia Publishing House; Philadelphia: Fortress Press, 1955–86), 33:28. Hereafter cited as LW.

[20]This is in contrast to Reformed dogmatics, where reason has long been a secondary *principium.* It holds to "human reason as the *principium quod* which draws the conclusions of faith from the unique *principium* of the infallible Scriptures." Gisbertus Voetius (d. 1676), quoted in: Henk Van Den Belt, *The Authority of Scripture in Reformed Theology: Truth and Trust* (Leiden: Brill, 2008), 168.

Chapter 5

Denial of Authority

The new scientific approach to knowledge started in an 18th-century intellectual movement commonly called the Enlightenment—an immensely slanted, doctrinaire term. Science deals with facts from man's point of view. It cannot compute divine intervention or causation—those are bracketed off. In science method reigns supreme.

The scientific approach to truth was applied to the documents and doctrines of Christianity. "The deeper appreciation of the role of the human authors in the composition of the books of the Bible, which dawned during the Enlightenment, put a question mark behind the claim that the Bible is God's Word." Inspiration was not denied so much as the very idea of a revelation became incomprehensible. "Post-Enlightenment skepticism concerning inspiration runs so deep that some have attempted to deny that the Church ever embraced so faulty a doctrine."[1] The Bible's authority was consequently reformulated to fit within the narrow bounds of scientific knowledge.

A proposition is a statement that can be translated, reformulated, and confessed in human words. The propositional property of truth says that essential meaning can be handled and reworked by man. The modern railing against propositions started as an attack on the Bible's

---

[1]William Lane Craig " 'Men Moved by the Holy Spirit Spoke from God' (2 Peter 1.21): A Middle Knowledge Perspective on Biblical Inspiration," in *Oxford Readings in Philosophical Theology: Providence, Scripture, and Resurrection,* vol. 2, ed. Michael Rea (New York: Oxford University Press, 2009; http://www.leaderu.com/offices/billcraig/docs/inspiration.html), 157.

propositional use as an authority. Eventually came the [post-]modernist idea that human language cannot contain truth in all its fullness. With different standards for determining truth, it is no wonder the fundamentalist and the modernist do not agree.

Not the content but the very concept of revelation was attacked. Divine truth was deemed incompatible with certain, real knowledge, because it was established by God's direct authority. The Enlightenment philosopher Baruch Spinoza "denied propositional revelation" as "worshipping paper and ink in the place of God's Word."[2] However, this charge of bibliolatry, the worship of the Bible, is not about worship, but the fact that no knowledge is allowed to be above and beyond the scope of human reason. This philosophical attack assumed a distinction between God's eternal Word and the tangible words of the Bible—the fundamental bias of modernism. One defender of the orthodox view of biblical inspiration wrote in 1828 that for "modern opposers of revelation . . . reason is their God."[3] Modernism, or "Enlightenmentism," is essentially the rebellion against any inerrant rule or standard outside of man.

Moderns are consumed with authority and its basis. But lots of talk about authority is just that: talk. Authority is a neutral word regarding knowledge. It does not imply correctness or rightness, as with governments and parents. To have the Bible as *an* authority is not to believe it is God's Word. Here the modern denial of authority found its focus in the only inerrant authority: God's communication.

"The traditional Christian doctrine of revelation is incompatible with the exercise of critical historical and scientific judgment."[4] New methods of verifying knowledge put God's interaction with the world on shaky ground. "Until recently, Scripture has been accepted as a form of revelation."[5] A key doubt of modernity is whether Scripture

---

[2] Norman L. Geisler in *Inerrancy*, ed. N. L. Geisler, 317.

[3] Thomas Hartwell Horne, *An Introduction to the Critical Study and Knowledge of the Holy Scriptures*, vol. 1 (London: T. Cadell, 1828; https://books.google.com/books?id=kXVAAAAAcAAJ), 21.

[4] William J. Abraham, *Divine Revelation and the Limits of Historical Criticism* (New York: Oxford University Press, 1982), 4.

[5] A. M. Fairweather, *The Word as Truth: A Critical Examination of the Christian Doctrine of Revelation in the Writings of Thomas Aquinas and Karl Barth* (London: Lutterworth Press, 1944), v.

is the highest authority and of the purest divinity, that is, God's own direct words. Knowledge itself has been redefined to exclude divinely established facts.

The very idea of an oracle—an utterance of God to man—has become foreign and unintelligible. In pre-modern times this passage was a noted bulwark for Scripture's authority: "The Jews were entrusted with the oracles of God" (Rom. 3:2). In medieval terms "prophetic speech" denoted exactly the same—oracular speech of the deity, not just unknowable or future information. The modern scientific mindset, however, cannot account for this.

A divine message that cannot be critiqued or weighed by modern-enabled reason does not fit with a scientific view of truth. Prophetic speech, as such, is "not explicable by natural or scientific laws."[6] In the pre-modern age "oracles, seers, theophanies, divine signs and wonders" were taken for granted and generally believed.[7] The legitimacy and interpretation of these revelations, including the Bible, were often in dispute, but not the fact that God could communicate and man's knowledge was incomplete without His revelation.

God has been distanced from man in every aspect of life, giving rise to the modern scientific mindset. Instead of crediting the divine for this world's operation, we give thanks to "natural laws" and purely materialistic causes. The doctrine of Scripture shows how far man has distanced himself from God in intellectual thought. The modern idolatry is actually a perversion of the Gospel of Christ. Enlightened man divinized himself, so no other gods or supreme authorities were allowed, visible or invisible.

> The prevalent apostasy under the later and more spiritual covenant was an apostasy of that nature which might have been expected from analogy. It was represented (in contradistinction to the grosser sin of idolatry among the Jews) to be a more 'spiritual and intellectual rejection of the Deity;' . . . as now revealed through the Gospel: which, of course, in-

---

[6]This is the modern definition of a "miracle." It is deist and unbiblical. Everything is miraculous that is influenced by God in the Bible, not just what cannot be rationally explained. Martin Luther puts basic 'daily bread,' that is eating and all daily happenings, in the supernatural category, so that thanks to God should be given for it. *Wikipedia, The Free Encyclopedia,* https://en.wikipedia.org/wiki/Miracle.

[7]Emil Brunner, *Revelation and Reason: The Christian Doctrine of Faith and Knowledge,* trans. Olive Trad Wyon (Philadelphia: Westminster Press, 1946), 4.

cludes a rejection, to a corresponding extent, in either case, of his existing special revelation.[8]

The question is not one of authority in general, for there are structures of authority everywhere. But are constitutions, the vagaries of tradition, and man's highest thoughts normed by direct divine authority? The authority of Scripture is the highest and most unconditional authority possible, but only if God spoke it and still speaks the same in it. God's words claim the highest authority, for they created the world in six days and still uphold it. Scripture connects God's many communications in various forms to Christ the Creator, who sustains all things:

> Long ago, at many times and in many ways, God spoke to our fathers by the prophets, but in these last days he has spoken to us by his Son, whom he appointed the heir of all things, through whom also he created the world. He is the radiance of the glory of God and the exact imprint of his nature, and he upholds the universe by the word of his power (Heb. 1:1–3).

A man-centered approach to knowledge redefined nature and then Scripture, leading to the invention of a new Christ. If "fundamentalism is a prison to escape from," Christian truth or anything that limits man's authority is the greatest evil.[9] In modernism intellectual and physical liberty came to symbolize evangelical freedom—in contrast to the freedom from God's wrath in the forgiveness of sins.

Why has inspiration been redefined, discarded, and made theologically inert? "For the first time in world history there is mass atheism and a completely secular culture."[10] There is no place for revelation that goes against the starting premise of modern truth. This has caused the critical doctrinal failure of not identifying Scripture with the Word of God. It has become impossible for the modern theologian to state clearly the full divine status of Scripture's words.

If Scripture is not God's Word in the most absolute sense, it is man's word. There is no divine status for what emanates from sinful, sanctified, or even pre-Fall man. Only God's authority is unqualified,

---

[8] John Miller, *The Divine Authority of Holy Scripture Asserted: From its Adaptation to the Real State of Human Nature, in Eight Sermons* (Oxford: W. Baxter, 1817; https://books.google.com/books?id=lllKAAAAYAAJ), 53.

[9] J. Barr, *Beyond Fundamentalism*, vii.

[10] E. Brunner, *Revelation and Reason*, 5.

whereas man's is always limited. What is called "fundamentalism" is not a basic Christian approach to truth: it is the basic pre-modern approach to truth. Divine communication is held to be the paramount knowledge. It is unquestionable, just as much as God Himself. Revelation requires full submission, even though it is unprovable and potentially irrational. Modernism is an intellectual atheism, a refusal to accept what is not convincing or attestable. It is a most satanic attack on Christ in these last days, with consequences in every area of life.

Chapter 6

Scientific Theology

The progress of science is modest. It has not improved man, though it has improved external things. Yet the biblical Christ did not come to improve this world. He came to bring us into a new heavens and a new earth, after the resurrection from the dead. Scientific study is a fine and godly vocation, but its method does not have a monopoly on truth. Rather, it should be theology which claims first place for those who follow Christ.

There is no getting around the scientific approach in all areas of knowledge, even if it is accepted uncritically. "Science explains the phenomena, pushing God (religion) or the gods (myth) aside."[1] Scripture is not valued as the oracles of God, but rather seen in negative terms. In 1891 the Scriptures were called "a yoke which [moderns]—unlike their fathers—are unwilling to bear."[2] This is a marked reversal of Basil of Caesarea's (d. 379) submission: "Shall I then prefer the foolish [human] wisdom to the oracles of the Holy Spirit?"[3]

The scientific approach to knowledge is not complete, but it has the look of completeness. " 'I do not know.' Such words are no longer proper

---

[1] S. J. Nichols and E. T. Brandt, *Ancient Word, Changing Worlds*, 20.

[2] Joseph Henry Thayer, quoted in: S. J. Nichols and E. T. Brandt, *Ancient Word, Changing Worlds*, 25.

[3] "Hexaemeron Homily," in *Nicene and Post-Nicene Fathers,* second series, vol. 8, ed. Philip Schaff (Buffalo, NY: Christian Literature Publishing, 1895; http://www.newadvent.org/fathers/32019.htm), IX:1.

among us modern people."[4] Nothing is out of bounds in a truly scientific investigation. So "it is uncritical to ascribe *a priori* [beforehand] God's authority to Scripture."[5] The issue is not the correctness of theological conclusions, but the very starting point and authorization for every conclusion. Modernist theology, if it is to be truly scientific in character, is centered on man and can only be spoken of with human authority.

Revelation is a modern word to encompass God's involvement in the scientific accounting of the world. Revelation, though, cannot be confirmed or refuted.[6] In the scientific model, Scripture must be proved verse by verse. That is a tall task, so the whole notion of inerrancy seems absurd. This is not because of any actual contradiction, but because inerrancy has been arrogated by modern man. Truth is bestowed on other objects only after the heaviest consideration and interrogation.

The English philosopher Francis Bacon (d. 1626) speaks of a radical, unrelenting criticism: "we should at once and with one blow set aside all sciences [received knowledge] and all authors; and that too, without calling in any of the ancients to our aid and support, but relying on our own strength."[7] This scientific criticism told theology there cannot be a divine book, and theology has obliged and built her castles on fashionable sands, only so the next theological genius can destroy them and build anew. A theology that is not timeless is not worthy of the name. It is mere human opinion.

Facts are now the exclusive domain of science, so religious authority and information have been ungrounded and severed from revelation. "Natural science has taken the place of ancient philosophy."[8] So theology is left with the scraps. Science is an all-consuming quest for factual knowledge based on man's viewpoint and certainty. Modern theology has given up indisputable facts and history, and therefore all significance. All has been given up for the respectable name of the modernist god: "science."

---

[4]F. Bettex, *The Bible the Word of God*, 7.

[5]Raymond Surburg, "Implications of the Historico-Critical Method in Interpreting the Old Testament: Part 1," *Springfielder* 26:1 (1962), 15.

[6]F. Bettex, *The Bible the Word of God*, 228.

[7]Quoted in: Christopher Hill, *Antichrist in Seventeenth-Century England* (New York: Oxford University Press, 1971), 37.

[8]Tom G. A. Hardt in *Hermann Sasse: A Man for Our Times?*, ed. John R. Stephenson (St. Louis: Concordia Publishing House, 1998), 156.

Pre-moderns upheld inspiration not as a theory to be tested, but an affirmation of the true author of Scripture, the Holy Ghost. While the word is still common, "the inspiration of the Bible [is] an unintelligible and antiquated notion." But "without revelation there is no possibility of true, absolute knowledge."[9] Theology assumed priority on the basis of revealed knowledge from a divine communication. "Because theology relied ultimately on revelation, . . . its knowledge and wisdom [transcended] that of the secular sciences."[10] Modernism is the product of a reversal of this pre-modern ordering.

Science as a man-centered activity is not inerrant, nor can it be a foundation for the truth. Its insights, which might be true as far as man can discover, pale before God's most authoritative Word. Man has been given authority and dominion over the things under him in creation. So science is thus a perfectly legitimate, though always limited, activity for studying physical objects. But to consider God as an object for man's theological laboratory is patently blasphemous.

---

[9] F. Bettex, *The Bible the Word of God*, 173, 39.
[10] Edward Grant, *God and Reason in the Middle Ages* (New York: Cambridge University Press, 2001), 208.

Part III

A Reverse Historical Comparison

Chapter 7

The Medieval Doctrine of Scripture

An errant Word of God was unthinkable until the 17th century.[1] "In the Medieval Age inspiration was axiomatic."[2] Revelation, in the abstract, was not to be questioned any more than God Himself. Although there were differences over the sufficiency, clarity, and interpretation of Scripture, the idea that theology was revealed was not in dispute. Natural, scientific facts were biblical facts, and no authority outside of revealed knowledge, in theory, had a privileged place.

The corollary to revealed knowledge is the deprecation of man's rational ability. "Give the Holy Ghost the honor of being wiser than yourself, for you should deal with Scripture that you believe that God Himself is speaking." Again Luther: "I must bring my own ideas into captivity and assent to the Word even if I do not understand it."[3] The facts of divine revelation were viewed as divine doctrines, which ruled as authoritatively as God himself. This explains why heretics were killed in the pre-modern era. The guilty had committed crimes against

---

[1] Robert D. Preus in *Inerrancy,* ed. N. L. Geisler, 357.

[2] Allan Andrew Zaun, "A Study of the Idea of the Verbal Inspiration of the Scriptures with Special Reference to the Reformers and Post-Reformation Thinkers of the Sixteenth and Seventeenth Centuries" (PhD diss., University of Edinburgh, 1937; https://www.era.lib.ed.ac.uk/handle/1842/10291), 117.

[3] Johann Michael Reu, *Luther and the Scriptures* (Columbus: Wartburg Press, 1944); reprint, *Springfielder* 24:3 (1960), 51, 52.

divine knowledge, which is above every human law. Revelation, God's Scripture, was assumed by most to be the foundation of all knowledge.

Luther's doctrine of inspiration is biblical, but it appears thoroughly medieval compared to theology of the modern period. It begins not with a critical investigation of Bible passages, but man's place before God: "Our reason cannot know the divine nature but still wants to judge concerning that about which it knows nothing."[4] Theology without a divine message is impossible.

In 1441, before Luther's birth, it was professed that there is "one and the same God as the author of the Old and New Testament, that is, of the Law and the Prophets and the Gospel, since the saints of both Testaments have spoken with the inspiration in the same Holy Spirit."[5] Scripture's authority was based on its unity, by virtue of the one divine Author. Inspiration was not a problem. It was an affirmation that the words of Scripture are the Spirit's words.

"Dictation theory" is a modern canard and verbal grenade, but the word "dictate" is the universal Christian word used to describe the result of inspiration, even if tradition in practice dominated:

> all books of the Old and of the New Testament—seeing that one God is the author of both—as also the said traditions, as well those appertaining to faith as to morals, as having been dictated, either by Christ's own word of mouth, or by the Holy Ghost. . . .

This confession of the Roman church, ostensibly correcting Lutherans, also uses the phrase "from the Apostles themselves, the Holy Spirit dictating [*Spiritu Sancto dictante*]."[6] Tradition here is seen as compatible with the Spirit's inspiration, because its origin is in the same God. "Dictation" is not a physiological description of the particular workings of the Spirit in the human author, but denotes that whatever the process, the result is exactly the same as if God dictated it directly. "The Holy Ghost inspires and dictates to them what they were to deliver of the mind of God . . . so that the very words of Scripture are to

---

[4] Quoted in: J. M. Reu, *Luther and the Scriptures*, 36.

[5] Pope Eugene IV, "*Cantate Domino*," Papal Bull from Council of Florence (http://catholicism.org/cantate-domino.html).

[6] Council of Trent, Fourth Session (1546). Philip Schaff, *Creeds of Christendom*, 3 vols. (Grand Rapids: Baker Books; rev. ed., 1984), 2:80.

be accounted the words of the Holy Ghost."[7] The modern offense is not in any description of inspiration itself, but the claim that God speaks in a supremely authoritative way. "The view has become almost universal that to-day we can no longer hold the mediæval conception of the inspiration of the Holy Scriptures."[8] Using popular Early Church imagery, this 1549 Saxon confession leaves no room for man to question a single word of Scripture:

> For although both, prophets and apostles, like ourselves, were natural men and descendants of Adam they, nevertheless, have not spoken nor written as men write out of their own mind; and their books and writings are certainty not their own human imaginings. It is God the Holy Ghost who spoke through their mouth and wrote through their hand. He is the real master and author of Holy Scripture, who to make known to men his word and teaching used mouth and hand of the prophets as His organ in no other way than the prophets used their pen and ink to write the word and as David used his harp to play on it.[9]

The medieval doctrine of Scripture is actually the position of Christ and of the whole Church until modern, atheistic times. There is no article on Scripture in the 16th-century Lutheran confessions, a fact that modern Lutherans delight in asserting. But given the context, we should no more expect an article on inspiration than one combating Darwin's much later evolutionary heresy. Inspiration was not an issue until after the scientific revolution in the 17th century. In any controversy, even against Satan himself, Jesus made the verdict on the basis of bare Scripture citations (Lk. 4:3–13). Like Jesus, "fundamentalists often display an amazing familiarity with Scripture."[10]

---

[7] Matthew Henry (d. 1714), quoted in: H. D. McDonald, *Theories of Revelation*, 1:201.

[8] F. Bettex, *The Bible the Word of God*, 170.

[9] This "Confession of the Landed Estates in Thuringia" was accepted by the Duchy of Saxony at Weimar. Quoted in: J. M. Reu, *Luther and the Scriptures*, 68. It was included in this local confession which was supplanted by the Book of Concord: "*Corpus Doctrinae Thuringicum* in Ducal Saxony, of 1570, containing the Three Ecumenical Symbols, Luther's Catechisms, the Smalcald Articles, the Confession of the Landed Estates in Thuringia (drawn up by Justus Menius in 1549), and the Prince of Saxony's Book of Confutation (*Konfutationsbuch*) of 1558." F. Bente, *Historical Introductions to the Symbolical Books of the Evangelical Lutheran Church* in *Concordia Triglotta: The Symbolical Books of the Evangelical Lutheran Church*, ed. F. Bente and W. H. T. Dau (St. Louis: Concordia Publishing House, 1921), 7.

[10] L. Jordahl, "The Theology of Franz Pieper," 131.

It is falsely claimed by modern Lutherans that Luther did not use the word "inspiration:" "Luther and the Confessions also do not use the term."[11] But a more careful scholar, using statistical analysis, rebuts that claim: "The theological concept 'inspiration' (and its derivatives [*inspirare and inspiratio* in Latin]) occurs 96 times in the writings of Luther," and most occurrences are "evenly spread throughout Luther's writings."[12]

Inspiration is an abstraction, a description of the process by which man's words are also God's Word. The result of inspiration, that the Bible *is* God's Word, is more significant than the historical fact of inspiration. But without the historical act, there is no connection between God the Holy Spirit and the words we possess today. A modern innovator claims that Lutherans have actually held the Reformed view of inspiration for the "last three hundred years": "This distinction between the primary [God's] and secondary [the human writer's] authorship was an invention of the seventeenth-century dogmaticians."[13] It is tragically ironic that such people "confess" the Nicene Creed, oblivious of its meaning, and deny the most clear historical facts:

> But who wrote these words [of the book of Job] is quite a pointless question when we believe confidently that the Holy Spirit is the true author of the book. The writer is the one who dictates things to be written. The writer is the one who inspires the book and recounts through the voice of the scribe the deeds we are to imitate.[14]

That aligns with this statement: "The Nicene Creed sets forth the whole doctrine of verbal inspiration in embryonic form."[15]

The pre-modern Luther used much stronger words than "verbal inspiration." "The Holy Spirit wanted to write this to teach us," so He is the true author and writer of Scripture.[16] No stronger view can be held than that the Spirit of Christ is the literary author of the words

---

[11] David P. Scaer, *The Apostolic Scriptures* (St. Louis: Concordia Publishing House, 1971), 16.

[12] M. Ruokanen, *Doctrina divinitus inspirata*, 49.

[13] P. H. Nafzger, *These Are Written*, 21.

[14] Gregory the Great (d. 604), *Commentary on the Book of Blessed Job* (http://faculty.georgetown.edu/jod/texts/moralia1.html), I.2.

[15] Siegbert Becker, *The Scriptures—Inspired of God* (Milwaukee: Northwestern Publishing House, 1971), 12.

[16] J. M. Reu, *Luther and the Scriptures*, 35.

of the Bible. It is claimed that strict verbal inspiration "emerged only over the last two centuries" "largely in reaction to the rise of historical-critical scholarship."[17] While the phrase "verbal inspiration" is new, it would have been redundant to pre-moderns. How would God the Spirit communicate non-verbally in *written* words or errantly in true, divine words?

What Christians of all ages previously accepted must now be spelled out in child-like detail: Scripture is written in words that are not false. All other effects of inspiration, besides the verbal, are irrelevant to the doctrine of Scripture. That God cannot lie and His written word is factual in all things, even where science makes claims, must now be defended. However, there is nothing new in this. While the language of verbal inspiration, inerrancy, and the correspondence theory of truth seem inert, this is precisely where God's doctrine is being attacked and must be defended.

The pre-modern confession of Scripture, however, is more clear and theologically astute than speaking of "verbal inerrancy." To be content with a confession of Scripture's inerrancy is utterly un-theological. A math proof can be without error, but it is not the most sure foundation for all divine knowledge. The opposite of modernism is not Christianity, though its heresies must be resisted by every true Christian.

The modern fight over inspiration is really about the possibility of divine communication. "Scripture *is* revelation," that all the human words of Scripture are God's words, is not a confession moderns can make.[18] Revelation is an abstraction, a mere anachronism, from the vantage point of scientific truth. But the Bible is not an idea we may toy with, but actual words God has spoken which remain eternally valid. "There is no other proper correlative for revelation than faith! To try to approach revelation as a thing, an object to be dissected and juggled, is to destroy the very thing of which one is seeking to get hold."[19] Inspiration, in the fullest sense, asserts that Scripture is divine

---

[17]Richard Gaillardetz, *By what Authority?: A Primer on Scripture, the Magisterium, and the Sense of the Faithful* (Collegeville, MN: Liturgical Press, 2003), 17.

[18]Gerhard Maier, *The End of the Historical-Critical Method*, trans. Edwin W. Leverenz and Rudolph F. Norden (St. Louis: Concordia Publishing House, 1977; reprint, Eugene, OR: Wipf and Stock, 2001), 63.

[19]Eugene F. Klug, "Review of *The End of the Historical-Critical Method*," Spring-

revelation. This "was recognized by Jews and Christians of every age."[20]

The fundamentalist position is in fact the universal view, outside of the Enlightenment-induced secularization of the West. The Roman church remained more steadfast in this teaching than most Protestants, resisting historical criticism until the 1940s. Until the scientific method of studying Scripture was adopted, the historic position was upheld: "because the Holy Ghost employed men as His instruments, we cannot therefore say that it was these inspired instruments who, perchance, have fallen into error, and not the primary author."[21] An erring Scripture would necessarily imply an erring God. "The rise of the historical critical research of the Bible also had its effect on Catholic theology so that in defining the concept of biblical inspiration the emphasis has biased to express the doctrinal inerrancy of the Scriptures."[22] The increased emphasis on the Bible's inerrancy has resulted from an equally potent modern attack on it.

Normative for Everything

In one of the first detailed confessions of Scripture by Lutherans, it is said that the Bible is "the word and mouth by which the eternal majesty of God reveals itself to the human race so that no man can perceive or know anything certain and trustworthy unless he perceives and learns it from Holy Scripture."[23] This is the basic medieval view: theology is the queen of the sciences, because of its divine basis. Luther said concerning Scripture: "This queen must rule, and everyone must obey, and be subject, to her."[24] The confession of its oracular nature as a divine communication meant detailed attention to the text and its grammar. Scripture's authority was unlimited. It asserted the truest facts in every area of knowledge. Theological truth judges every other

---

fielder 38:4 (1975), 291.

[20] T. H. Horne, *An Introduction to the Critical Study and Knowledge of the Holy Scriptures*, 220.

[21] (1893), Pope Leo XIII, "*Providentissimus Deus*: On the Doctrine of the Modernists," 20.

[22] M. Ruokanen, *Doctrina divinitus inspirata*, 139.

[23] For historical details see n. 9 in this chapter. Quoted in: J. M. Reu, *Luther and the Scriptures*, 68.

[24] *Galatians Commentary* (1535), LW 26:58.

truth.[25]

The modernist–fundamentalist debate is in essence a "battle over the unrestricted authority and truth of scripture." Modernism is an "intellectual process which ended with the dethroning of the Bible as the authoritative source of all human knowledge and understanding."[26] While Luther is often posited as the first modern, on the primary basis for knowledge, he was thoroughly medieval. Counter to that, a modern Lutheran claims: "Unlike many proponents of the doctrine of inspiration, Luther did not try to resolve apparent discrepancies that occasionally appear in his reading of the Scriptures."[27] The facts do not corroborate this assertion.

It is the historical reservation epitomized in historical criticism which in the modern era causes so much mischief. No true modern could ever say with Luther: "That is a small matter if the whole calculation [of the age of the world] is otherwise certain and a doubt remains in only two or four years."[28] Luther, late in life, wrote an entire work calculating the age of the world from Scripture, entitled "Computation of the Years of the World."[29] Though not without difficulties, Luther says: "I have carefully reckoned together the years of world," based on Scripture.[30] "It is highly necessary for all men to know the ordering of the years from the beginning of the world . . . ." What was his basis? "This cause has moved me, while not despising the historians, to prefer Holy Scripture to them. . . . For I believe that in Scripture God speaks, who is true . . . ."[31] On Is. 9:2, Luther comments: "The sun has already been shedding its light 5545 years, but it does not illuminate a single man unto eternal life."[32] Very few fundamentalists today would be so bold!

---

[25] Klaus Scholder, *The Birth of Modern Critical Theology: Origins and Problems of Biblical Criticism in the Seventeenth Century,* trans. John Bowden (London: SCM Press, 1990), 106.

[26] K. Scholder, *Birth of Modern Critical Theology,* 7–8.

[27] P. H. Nafzger, *These Are Written,* 110.

[28] Quoted in: K. Scholder, *Birth of Modern Critical Theology,* 77.

[29] *Supputatio annorum* (1541). K. Scholder, *Birth of Modern Critical Theology,* 68; Kurt E. Marquart, *Marquart's Works,* 10 vols., ed. Herman J. Otten (New Haven, MO: Lutheran News, 2014–15), 8:22.

[30] Quoted in: J. M. Reu, *Luther and the Scriptures,* 52.

[31] Luther, quoted in: K. Scholder, *Birth of Modern Critical Theology,* 71, 68.

[32] (1543), *Bethlehem and Calvary: The Christmas and Easter Book of Dr. Martin Luther: A Thorough Exposition of Chapters 9 and 53 of the Book of the Prophet*

"In his interpretation of history, Melanchthon agreed with Luther, both Reformers holding that the book of Daniel provided basic clues to the understanding of the past."[33] Scripture was not a problem for them; it was the only lens for divine understanding. Melanchthon quotes Joshua 10 in his physics book.[34] The later Lutheran Aegidius Hunnius (d. 1603) summarized the standard Lutheran position on the Scriptures: "it is a sacrilege and a crime to dispute their veracity in the minutest detail."[35] This was not lip service. The Bible was actually used as the divine key to all knowledge.

It is not scientific truth that endangers Christianity, but false conclusions and methodology applied outside their proper purview. In fact, it is not natural scientists, nor the Enlightenment philosophers whom are first credited with a rational approach to Scripture's teachings, but a small, heretical sect of the late 1500s. Socinianism "is the origin and embodiment of all modernist heresies" because "the intrinsic connection between the dogmas of Christianity was investigated before the forum of [human] reason."[36] Faustus Socinus (d. 1604) founded an antitrinitarian Polish religious group on rational principles. A new method of approaching Scripture, based on a new doctrine of man, changed everything. Scripture was freed from all prejudice of a doctrinal nature. "Socinian criticism was aimed less at the content of dogma than its presuppositions." This new denial of any limit to human reason is "the modern understanding of reality."[37]

Isaiah, trans. Kenneth K. Miller ([K. K. Miller], 1988), 16.

[33] Ulrich Michael Kremer in *The Seven-headed Luther: Essays in Commemoration of a Quincentenary,* ed. Peter Newman Brooks (Oxford: Clarendon Press, 1983), 209.

[34] "Lutheran theologians passionately defended this scriptural proof [Josh. 10] for more than a century and probably regarded it as central, even if for understandable reasons it is no longer so for a Lutheran systematic theologian in the twentieth century." K. Scholder, *Birth of Modern Critical Theology,* 49, 155.

[35] Quoted in: Robert D. Preus, *Inspiration of Scripture: A Study of the Theology of Seventeenth Century Lutheran Dogmaticians* (Mankato, MN: Lutheran Synod Book, 1955), 80.

[36] K. Scholder, *Birth of Modern Critical Theology,* 29; Wilhelm Dilthey, quoted in: K. Scholder, *Birth of Modern Critical Theology,* 30.

[37] K. Scholder, *Birth of Modern Critical Theology,* 28, 39.

Chapter 8

The Spirit is the Primary Author

The words of the Bible are entirely and fully the Spirit's words. This was the settled conviction of Luther: "He who wants to hear God speak should read Holy Scripture."[1] This view allows the heaviest emphasis on a single word. Marin Chemnitz, a generation after Luther, spoke of Scripture overturning his opponents who deny the Lord's Supper: "One word of the Holy Spirit overturns the wisdom of these men and makes it foolishness. Their learning becomes boorish and their hypocrisy an abomination."[2] This was an all or nothing view: "For whoever despises a single Word of God does not regard any as important."[3]

Why such bravado in confessing a Scripture they did not fully understand? It was a confession of Christ, whom they saw as the speaker, rather than a human judgment of the Bible's content. "The highest proof of Scripture derives in general from the fact that God in person speaks in it."[4] This was the near unanimous position of Christians until

---

[1] *Against the Roman Papacy, An Institution of the Devil* (1545), LW 41:332.

[2] Martin Chemnitz, *The Two Natures in Christ,* trans. Jacob A. O. Preus (St. Louis: Concordia Publishing House, 1971), 23.

[3] Luther, *1 Timothy* (1528), quoted in: *Inerrancy,* ed. N. L. Geisler, 380.

[4] This statement is in the section entitled: "The witness of the Holy Spirit: this is stronger than all proof." John Calvin, *Institutes of the Christian Religion,* 2 vols., The Library of Christian Classics, ed. John T. McNeill, trans. Ford Lewis Battles (Philadelphia: Westminster Press, 1960), I.vii.4, 78.

well past the Reformation. The poet John Milton (d. 1674) referenced God the Spirit as the Scripture's authority, because He is ultimately the origin of its words: "The Spirit is the authority of Scripture."[5]

The traditional view of inspiration is quite simple: "God spoke to our fathers by the prophets" (Heb. 1:1). The human authors are affirmed, but they are not what made Scripture unique. "The Holy Spirit is the simplest author and adviser in heaven and on earth."[6] The Bible's dual authorship, divine and human, was not problematic. Luther held that Scripture is the eternal Word of God.[7] Authority comes undiluted from the Spirit who speaks all the words recorded. "Of all rules it is safest and most correct to speak with Scripture itself and to imitate the language of the Holy Ghost."[8]

The modern split between inspiration and revelation completely changes the nature of theology. For Thomas Aquinas (d. 1274) there is no inspiration without revelation. Inspiration is simply a subset of prophetic speech, under the heading of "Prophetic Revelation." The things inspired by the Spirit are "equivalent to the content of the 'revelation' to be communicated."[9] Aquinas is critiqued mercilessly as a rationalist, but his position is fundamentalist today. He had no problem with God speaking in a book. "Aquinas never asked the question of how divine and human actions had jointly produced the Bible."[10] He saw no philosophical contradiction between God who produced, and still authorizes, the Bible and its human authors, as moderns do.

Inspiration is a miracle that cannot be explained by human reason. It is the scriptural fact that men and the Spirit are fully the Bible's authors. The traditional concept of accommodation affirms that the "use of human words . . . in no way implies the loss of truth or the lessening of scriptural authority."[11] Historical circumstances and human

---

[5] Quoted in: A. A. Zaun, "A Study of the Idea of the Verbal Inspiration," i.

[6] *Answer to the Hyperchristian, Hyperspiritual, Hyperlearned Book by Goat Emser in Leipzig* (1521), LW 39:178.

[7] K. Scholder, *Birth of Modern Critical Theology*, 21.

[8] M. Chemnitz, *The Two Natures in Christ*, 395.

[9] Pierre Benoit, *Aspects of Biblical Inspiration,* trans. J. Murphy O'Conner (Chicago: Priory Press, 1965), 44, 51.

[10] Denis M. Farkasfalvy, *Inspiration and Interpretation: A Theological Introduction to Sacred Scripture* (Washington, D.C.: Catholic University of America Press, 2010), 149.

[11] Richard A. Muller, quoted in: Hoon Lee, "Accommodation: Orthodox, Socinian,

authorship do not diminish the Spirit's authority and truthfulness in the least. Without this miracle of inspiration, the "Bible is the Word of God in a *derivative* or *figurative* sense."[12] "All we have [in Scripture] is human speaking about God's speaking," therefore dogmatics is "impossible."[13] This is quite a different spirit from that of Clement of Rome (d. A. D. 99), who is not a post-Reformation scholastic by all accounts: "He (Christ) Himself by the Holy Ghost thus addresses us [in Scripture]."[14]

The human authors were elevated and given divine words, so that they did not speak merely a human word, but a fully divine one. This is no difficulty for the Holy Spirit, who also conceived the God-man Jesus in Mary's womb. "In medieval times the nouns *calamus* [pen] and *secretarius* [secretary] and the verb *dictare* [to dictate] are often found" in connection with inspiration.[15] This thinking goes back to the earliest days of Christianity. Hippolytus of Rome (d. 235) describes inspiration:

> For these fathers were furnished with the Spirit, and largely honored by the Word Himself; and just as it is with instruments of music, so had they the Word always, like the plectrum [pick], in union with them, and when moved by Him the prophets announced what God willed. For they spake not of their own power (let there be no mistake as to that), neither did they declare what pleased themselves.[16]

This musical instrument analogy "does, in fact, succeed in emphasizing both the divine and human aspects of Scripture, since the type of instrument selected by the musician will determine the character of the musical sounds produced by his playing." [17]

Dictation-language and the musical analogy allow God the Spirit to receive full credit as the author of Scripture. This language is the most catholic and Christian way to speak of inspiration. While to moderns it minimizes man, it is just an analogy, not a full accounting of inspiration.

> To designate this relation of instrumentality, the Church Fathers and also our old Lutheran theologians call the holy

---

and Contemporary," *Westminster Theological Journal* 75 (2013), 336.

[12] S. Becker, *The Scriptures—Inspired of God*, 18.

[13] A. M. Fairweather, *The Word as Truth*, 41.

[14] On Ps. 34. Quoted in: *Inerrancy and the Church*, ed. J. D. Hannah, 10.

[15] J. M. Reu, *Luther and the Scriptures*, 62.

[16] *De Christo et antichristo*, quoted in: W. L. Craig "Men Moved by the Holy Spirit Spoke from God," 160.

[17] W. L. Craig "Men Moved by the Holy Spirit Spoke from God," 158.

> writers penmen of the Holy Ghost, or His recorders, notaries, scribes, amanuenses [secretaries], and the like. . . . they are truly scriptural as long as the point of comparison is kept in mind.[18]

Scripture asserts that God and man are both fully its authors. Which one, divine or human, impacts theology and knowledge the most? The divine Author is the ultimate authority, since all authority is from Him (Rom. 13:1). It is not the secondary, human authors that make Scripture authoritative or useful for Christians today. Luther, while not using current terminology, could matter-of-factly say: "We must, therefore, believe that the Holy Spirit Himself composed this psalm and laid it out before us in order to instruct us regarding our sad condition."[19]

> The language of the Fathers both in the East and in the West, as well as their habitual handling of the Scripture, leaves little doubt that for many if not most of them God was, altogether simplistically, the literary author of the Bible. He had, through men, "written" the Biblical work; He had "dictated" it.[20]

The word *dictatio* [dictation], though a great offense to moderns, "denotes no more than a divine supplying or furnishing of words in Scripture," not precisely how that took place.[21] This is nothing but the simple teaching of the Nicene Creed, in which "we attribute to the Holy Spirit all of Holy Scripture and the external Word and the sacraments, which touch and move our external ears and other senses."[22]

What about tradition? Did the existence of an extra-biblical authority contradict the Bible's inspiration? No, it was a denial of the sufficiency of Scripture, that it contained all necessary knowledge. But tradition was seen as fully divine and revealed as Scripture. Since in the Roman "double-source theory" the Spirit was the author of both truths, no conflict was assumed.[23] The words from Christ's mouth passed down orally and the Spirit's inspiration were theologically equivalent, since

---

[18]Francis Pieper, *What is Christianity? And Other Essays,* trans. John Theodore Mueller (St. Louis: Concordia Publishing House, 1933; reprint, Malone, TX: Repristination Press, 1997), 242.

[19]Commentary on Ps. 90. LW 13:81.

[20]W. L. Craig "Men Moved by the Holy Spirit Spoke from God," 160.

[21]R. D. Preus, *Inspiration of Scripture,* 195.

[22]Luther, *On the Last Words of David* (1543), LW 15:275.

[23]R. D. Preus, *Inspiration of Scripture,* 133.

they had the same origin. The Reformers saw both as one and the same, found only in Scripture itself. The discomfort with God being able to speak was not a pre-modern concern, though Scripture in practice was frequently treated poorly. The Christian confession is that the Spirit's words are Christ's words. They are not two separate Gods, but one God with the Father.

It is a paradox that God is the primary author of Scripture's words and thoughts, yet "God's spokesmen wrote willingly, consciously, spontaneously, and from the deepest personal spiritual conviction and experience."[24] This mystery, revealed by Christ, can only be believed, not explained or grasped by the human mind. Very early Irenaeus (d. 202) called the Scriptures "the Oracles of God, and Writings dictated by His Word and Spirit."[25] What is falsely called post-Reformation scholasticism and 19th-century Reformed thinking, the direct inspiration of Scripture by the Spirit, is actually the universal Christian conviction. The supposedly Lutheran scholastic dogmaticians make moderns seem utterly pagan in their description of God's written Word: "There can be no prophetic Word apart from the personal Word. How often do the Lutherans speak of Christ as the authority of Scripture and of Scripture as Christ's Word!"[26] Inspiration—the Spirit of Christ's own revealing—set firm limits to true knowledge for Luther: "what God does not want revealed, that I am not to know. . . . If I am not to know it, I should be quiet—or else I will break my neck."[27]

---

[24]Robert D. Preus, *The Theology of Post-Reformation Lutheranism: A Study of Theological Prolegomena*, 2 vols. (St. Louis: Concordia Publishing House, 1970), 1:291.

[25]Quoted in: T. H. Horne, *An Introduction to the Critical Study and Knowledge of the Holy Scriptures*, 81.

[26]R. D. Preus, *Theology of Post-Reformation Lutheranism*, 1:269.

[27]Quoted in: Siegbert W. Becker, *The Foolishness of God: The Place of Reason in the Theology of Martin Luther,* ed. John A. Trapp (Milwaukee: Northwestern Publishing House; 2nd ed., 1997), 15.

Part IV

The Scientific Subjugation of History

Chapter 9

A New Methodology for Knowledge

The scientific redefinition of all knowledge required a new basis for theology. The controversy of the new scientific worldview eventually undermined for many "the credibility of Scripture and thus the authority of God Himself."[1] Whereas Scripture formerly reigned and Aristotle was practically seen as inspired, a reversal occurred. What was aged became suspect to self-labeled "moderns."

The conflict, of course, would entangle God's own authority in Scripture. Today modernists claim that an inerrant Scripture is absurd and illogical. They operate with a different method of truth verification and a different conception of God. "Because the old [revealed] beliefs were unsettled and natural science offered a new criterion of certainty, Christianity felt the impact of science through its whole frame."[2]

"Science is more than a body of knowledge; it is a way of thinking."[3] This instrument for knowing became the method for discovering all truth. An "emphasis away from the data of special revelation toward the data of natural and physical science" led to "a new religion of

---

[1] K. Scholder, *Birth of Modern Critical Theology*, 46.
[2] Richard S. Westfall, *Science and Religion in Seventeenth-Century England* (New Haven, CT: Yale University Press, 1958), 10.
[3] Carl Sagan, *The Demon-Haunted World: Science as a Candle in the Dark* (New York: Random House, 1995), 25.

nature and reason."⁴ What cannot be verified by man—actual divine speech—cannot be considered true knowledge to one carefully following scientific methodology. It is ruled out from the start, by the principle of doubt. "One of the greatest commandments of science is to distrust arguments from authority."⁵ Science is methodology, not a storehouse of static facts that are accumulated.

"Knowledge was not a problem for the ruling philosophy of the Middle Age."⁶ It was taken for granted, because it had a divine basis: revelation. "Down to the close of the Middle Ages . . . questions of fact were settled not by experiment but by authority."⁷ An oracle from God can only be verified by God Himself. The critic of the Scriptures, the very oracles of God, approaches them as human writings—denying their authority completely. This was not seen as an objective method by pre-moderns.

The modern critic, however, reserves the right of private judgment. It would be unscientific, in principle, to exempt Scripture from an examination, based on all the available (extra-scriptural) evidence. The scientist Carl Sagan asks: "What sermons even-handily examine the God hypothesis?"⁸ The real question is: does the scientific approach have limits, or can it deal with all things? The scientific method has been used to redefine truth. Only what is within its limits of study and verified by its procedure can be deemed certain knowledge by moderns.

In science the basis of certainty is man the observer. Any religious truths must be verified by man to be factually true. This conflicts immediately with revealed truth, which is supernatural, and therefore unverifiable. It cannot be tested, only accepted or denied. Only a natural religion, available to all men, can survive a scientifically defined truth.

In adopting this method, criticism becomes the ultimate right and essential human characteristic. But in order for this to work, the whole supernatural realm must be redefined to make room for man in his

---

⁴Bruce Demarest in *Challenges to Inerrancy,* eds. G. R. Lewis and B. A. Demarest, 13.

⁵C. Sagan, *The Demon-Haunted World,* 28.

⁶Edwin A. Burtt, *The Metaphysical Foundations of Modern Physical Science: A Historical and Critical Essay* (London: Routledge and Kegan Paul; 2nd rev. ed., 1932), 2.

⁷Alan Richardson, *The Bible in the Age of Science* (London: SCM Press, 1961), 9.

⁸C. Sagan, *The Demon-Haunted World,* 34.

big, rationalistic britches. "It is the prerogative of reason to judge the credibility of a revelation."[9] Scientific truth, as defined by its method, reformulates the truth it purports to find. Direct revelation is ruled incompatible with man's knowledge—certain, scientific knowledge.

Man-made criteria are not the proper measure of Scripture. To treat the Scriptures as human documents, testable by natural methods, reduces them to merely human writings. "The fact that so little findings of modern science are prefigured in Scripture casts further doubt on the inspiration of Scripture."[10] A truth that is assumed to be naturalistic, since there is a physical limit to what man can examine, will only verify what he can confirm actually exists. But a special revelation that merely confirms general revelation would be a strange *special* communication. It would say nothing that man could not discover for himself.

Claims to objectivity notwithstanding, modern science has a definite beginning and philosophical heritage. In the early 1600s "[René] Descartes began his philosophical pilgrimage toward certainty by doubting everything except his own thinking."[11] This principle of doubt is the main presupposition of the scientific method. Its foundation is the human observer as the sole authority and guarantor of truth. No reliance on traditional authorities is allowed, hence every supposed external authority is relativized instantly. Everything is doubted, so that in scientism all facts are confirmed by empirical observation. What is to be certain, must first be doubted and tested critically. This is an entirely "new principle of knowledge."[12]

Within certain fields of study the scientific method is harmless. Man can doubt the things below him in creation, because authority over them is granted by God. But the new "scientific method was enthroned as the supreme way of knowing."[13] It became the avenue to a natural religion, making salvation available to all. Eventually the reach of scientism created a new definition of reality, positing "the historical character of existence" and "the time-conditioned character of each event and of our

---

[9] Charles Hodge, quoted in: *Inerrancy and the Church*, ed. J. D. Hannah, 259.

[10] C. Sagan, *The Demon-Haunted World*, 35.

[11] Richard A. Muller, *Post-Reformation Reformed Dogmatics: Prolegomena to Theology*, vol. 1 (Grand Rapids: Baker Books, 1987), 242.

[12] K. Scholder, *Birth of Modern Critical Theology*, 62.

[13] Carl F. H. Henry, "Narrative Theology: An Evangelical Appraisal," *Trinity Journal* 8:1 (1987), 3.

Scientific History: The Subjugation of History

distance from it in history."[14] This was done in the name of certainty.

It is remarkable that the first scientists themselves did not see a conflict between scientific study and religion. Rather, it was non-scientists, in their unscientific approval or traditional overreaction, who radically tried to redefine truth scientifically. Even Galileo said in medieval fashion that Scripture was "dictated by the Holy Spirit."[15] Yet there was a philosophical movement toward knowledge which is based on nature, proved by rational methods, over unverifiable revealed knowledge.

For moderns, "science is the true model of the world."[16] Anything under its authoritative umbrella is seen as ultimate truth. When facts are by definition established by man, revelation becomes outmoded and obsolete: "Verbal dictation, inerrant manuscripts, uniformity of doctrine between 1000 B. C. and 70 A D.—all such ideas have become incredible in face of the facts."[17] This new *method* of obtaining facts (truth) meant "the replacement of the biblical picture of history by a new picture of the real course of world history, based on experience and evidence."[18] What is "real" to the modernist must be obtained by this method, that is, by man himself.

During the emergence of modernism, it was not the results, but the new basis for supreme truth that was most radical. This statement from 1646 is definitely pre-modern: "Sure[ly] one would think that astronomical truths had more need of Scriptural confirmation than the Scripture of them."[19] The conflict is one of priority: does science back the teaching of Scripture or does Scripture have the right to criticize scientific results? For moderns, Scripture must bow to scientific methodology as the final and ultimate criterion of truth.

That reason, the unstated driver of the scientific method, is above any divine communication is the exact claim the Enlightenment thinker

---

[14]Gerhard Ebeling, *Word and Faith,* trans. James W. Leitch (Philadelphia: Fortress Press, 1963), 46.

[15]Quoted in: K. Scholder, *Birth of Modern Critical Theology,* 59.

[16]Norman L. Geisler in *Biblical Errancy: An Analysis of its Philosophical Roots,* ed. Norman L. Geisler (Eugene, OR: Wipf and Stock, 1981), 12.

[17]S. J. Nichols and E. T. Brandt, *Ancient Word, Changing Worlds,* 54.

[18]K. Scholder, *Birth of Modern Critical Theology,* 65.

[19]Alexander Ross (d. 1654), quoted in: R. S. Westfall, *Science and Religion in Seventeenth-Century England,* 21.

John Locke (d. 1704) made: "In all things . . . reason is the proper judge; and Revelation, though it may in consenting with [reason] confirm [reason's] dictates, yet [Scripture] cannot in such cases invalidate [reason's] decrees."[20] Revelation needs rational approval, even when it is believed. Without this "proof," it must be doubted according to the scientific principle. In the classical understanding a revelation that must first be judged and tested by man is no revelation at all.

Reason, in the "enlightened" method, is elevated to a god-like position. It always has privileged status above its object of study. If reason can judge God's direct communication, it judges God as a superior. This is the "neutral" method of science, applied to something decidedly non-neutral. The right to judge all things and question any authority is a hallmark of modernity. This method, enshrined by the social status of science, elevates man as critic, and therefore lord of all. It has become a universal religion: the right to criticize all things and discover truth for oneself.

Even the average "Bible-believer" of today seeks to make Scripture acceptable to science, and very rarely the reverse. Here doctrinal conclusions are not the problem. The foundation for truth claims, including religious doctrine, has changed. While Thomas the apostle seems very scientific in needing to verify with his own finger Jesus' wounds, the Lord simply calls him unbelieving. What is most fascinating about this account is that though he was invited to seek physical proof of Jesus' resurrection, the Word of physically present Jesus was sufficient for his return to faith. Thomas did not carry through in his bold task to objectively examine and pass judgment on the Lord's resurrection, when the spoken Word caused him to repent of his "objective" unbelief. "Jesus said to him, 'Have you believed because you have seen me? Blessed are those who have not seen and yet have believed'." (Jn. 20:24–29).

Authority became uncompelling to the scientific-minded. This new approach to obtaining knowledge *began* with "the rejection of all arguments based solely on tradition and authority."[21] Scripture gave way to rational proof as having the connotation of authority. Submission to

---

[20] John Locke, *An Essay Concerning Humane Understanding,* vol. 2, eds. Steve Harris and David Widger (http://www.gutenberg.org/files/10616/10616.txt, 2004), II.IV.XVIII:6.

[21] K. Scholder, *Birth of Modern Critical Theology,* 63.

any authority came to be seen as competing against factual knowledge.

Scripture to the modern man is idolatrous, because it does not admit objective investigation. Revelation appears as the ultimate legalism against the new gospel of scientific knowledge. "Reason was no longer prepared to submit to the authority of Scripture and tradition."[22] Based on perceptible discoveries and an increasingly scientific depiction of the world, the "old unity of world-view and faith was broken."[23] The search for ultimate truth centered on the new method for obtaining factual knowledge. The very word "worldview" denotes a working hypothesis open to criticism, not a foundational *a priori*. Man, in true scientific fashion, reserves the right to select a new perspective or "worldview."

René Descartes formulated a philosophical basis for the logical application of doubt in the pre-scientific decision. "The significance of Cartesian philosophy lies in its principle and not in its result." Confidence in reason led to an individualistic approach to certainty, that is, an entirely new way of thinking. Reason itself authorized the "independence of scientific research from all philosophical and theological principles."[24] Revelation became more and more questionable, as Scripture was rightly seen as incompatible with the naturalistic grounding of knowledge.

Natural knowledge and rational explanations took precedent over bare authority in the modern social climate. The visible fruits of the scientific mindset seemed to establish its truth over the unseen and intolerant truths of religious orthodoxies. "Advances in medicine, astronomy, physics, and the other natural sciences exercised influence on the minds and lives of the age."[25] It became difficult to think without the critical methodology first given prominence in the hard sciences. In the early 1800s "the claim was being made that no province of the universe could be shut off from scientific investigation."[26] The modernist mindset was born. It judges any supposed timeless orthodoxy by the strident principle of doubt and man's evolved enlightenment, not its supposed revelation.

---

[22] K. Scholder, *Birth of Modern Critical Theology*, 89.
[23] K. Scholder, *Birth of Modern Critical Theology*, 91.
[24] K. Scholder, *Birth of Modern Critical Theology*, 111, 63.
[25] Kenneth G. Appold in *The Bible in the History of the Lutheran Church: The Pieper Lectures,* ed. John Maxfield (St. Louis: Concordia Historical Institute, 2005), 23.
[26] H. D. McDonald, *Theories of Revelation*, 1:117.

A New Methodology for Knowledge

Under the authoritative pale of science, biblical study morphed into a scholarly activity for the professional exegete, rather than a religious activity done in the Spirit. Only the most trustworthy facts, as verified by the new principle of doubt, were allowed to establish the basis of religion. In short, a new religion, with man at the forefront, emerged from a synthesis of naturalistic deism and scientific biblicism. Under the name of "science," doubt become scholarly and religious.

Chapter 10

The Sordid History of Historical Consciousness

The word "old" was not pejorative until the modern era. History was something to respect, not criticize mercilessly. Age carried an air of authority. Before the Enlightenment pre-moderns "lacked what we call today a specific 'historical consciousness,' an awareness of the complex and continuous development that characterizes human life and culture . . . which generally permeates our understanding of human existence and thought." In the face of the new scientific truth, theology either resists scientific claims, or completely accepts them and denies the relevance of factual truth for religious knowledge. The latter defines modern theology, which accepts scientific truth as foundational. Fully "fashionable [are] affirmations of modern science and history while adhering to all articles of faith."[1] Christianity itself has been made a mental idol by the theologians tasked with teaching it.

"At the end of the seventeenth century, the issue of contradiction between the biblical description of the physical universe and the rapidly evolving scientific discoveries," caused the Bible to be treated as any other source material.[2] As a human document, it was in principle open to criticism. The error was not a theological one or merely a deviant interpretation of Scripture. Rather, secular knowledge overtook the

---

[1] D. M. Farkasfalvy, *Inspiration and Interpretation*, 122, 165.
[2] D. M. Farkasfalvy, *Inspiration and Interpretation*, 162.

divine. "Almost all the objections were argued from the side of science and history."[3] After all, how does man verify a perfect, divine book? He cannot because there is no point of comparison.

History itself seems straightforward, but has been redefined by moderns to exclude any authority above man. Man was made the center of history and critical lord of all things. "Historical-critical theology . . . takes the modern understanding of reality into account."[4] Submission to an authority as a pre-condition for historical study is out of the question. It is against the scientific approach, which is centered on human judgment. "Historicism discounts bias, prejudice and any interpretive filter" with "serene confidence in reason."[5] Divine authority is ruled out from the start. God may not speak with absolute authority when His words *must* be tested by the method of persistent and boundless doubt. Orthodox dogmaticians could only see "arrogance, ignorance or atheism" in this consistent criticism.[6]

Historical criticism embodies a very sophisticated view of history, but one that is mostly an unstated assumption. "Critical history is scientific history," which "involves a critical attitude toward the sources."[7] No one witness is to be trusted fully, so they are all to be equally and neutrally doubted. All alike must submit to the critic, who acts as the highest judge to confer authority. Moderns trace the development of ideas and think that once the human origin of an idea is stated, it has been fully explained. In this closed system, God is ruled out beforehand, or at the very least, marginalized. "The Bible was no longer the criterion for the writing of history; rather [critical] history had become the criterion for understanding the Bible."[8] This new arrogance will not allow a revealed accounting of history, which Augustine held: "whatever in secular histories runs counter to [Scripture] we do not hesitate to brand as wholly false."[9]

---

[3] D. M. Farkasfalvy, *Inspiration and Interpretation*, 165.

[4] K. Scholder, *Birth of Modern Critical Theology*, 3.

[5] Kenneth Hagen, *Foundations of Theology in the Continental Reformation: Questions of Authority* (Milwaukee: Marquette University Press, 1974), 6–7.

[6] K. Scholder, *Birth of Modern Critical Theology*, 5.

[7] A. Richardson, *The Bible in the Age of Science*, 54.

[8] Edgar Krentz, *The Historical-Critical Method,* in Guide to Biblical Scholarship: New Testament Guides, ed. Dan O. Via, Jr. (Philadelphia: Fortress Press, 1975), 30.

[9] *The City of God* 18.1, quoted in: *Inerrancy and the Church*, ed. J. D. Hannah, 59.

Under the assumptions of historical distancing and progress, all words were distanced from the present, even though real authority is expressed in words. The modern time period (the present critical bar) was elevated above the past. The critic gained his judging authority at the expense of all previous authorities. Man declared himself above all history, that is, over anything that would deny authority to his present critique. When "all knowledge (or even truth) is historically conditioned" this "allows history to be scientific."[10] This historical consciousness, or temporary atheism, relativized everything in the name of progress. By declaring himself enlightened, man threw off the old authorities, especially the noxious doctrine of sin which had shackled pre-modern man. Sin, as a limitation of the critic, would prevent objectivity in judging truth.

God was displaced in history, and history was instead centered on the reflecting consciousness of man. The old authorities could not apply to moderns, having been written for an unenlightened people. By insulating modern man from past authorities, modern history sees the past as bound to a past people. " 'Historical consciousness' . . . tells us that the mentality and experience of former generations, especially in other cultures, was so different from our own as to be largely incomprehensible to us."[11] Therefore, we have "the Bible's time-conditioned words," which speak only to specific historical situations, as with any other supposed authorities.[12] This has become the methodology of truth and a deeply ingrained cultural pattern. Science, though, is so limited in scope that it cannot offer certain truth or fulfill the heart of man. Modernism has excelled in materialism and trivial technologies, but in morals, meaning, and art it has produced much that is corrosive, meaningless, and inane.

God, who is older than all creation and outside of time, is not easily massaged into the naturalistic truth method. Can Scripture be read without a privileged status—treated like every other human writing? It can, but can this skeptical approach be called Christian? This method to achieve certainty is not a series of analytic steps, but a new approach to knowledge. The unstated question is *who* verifies facts—all facts, including the historical? Since history is not repeatable or directly observable, like the objects of hard science, a problem arises. How can

---

[10] E. Krentz, *The Historical-Critical Method*, 56.
[11] J. I. Packer in *Inerrancy and the Church*, ed. J. D. Hannah, 171.
[12] E. Krentz, *The Historical-Critical Method*, 62.

reason verify history, without complete or objective access to it? It can only do so by severing all ties to past conceptions of authority and centering the factual world on the critic.

This new approach or method submits everything historical to the rational observer. The historian "deals with all historic and literary phenomena of the past with the same method, *viz.*—the critical historical method."[13] Behind it is an arrogance of superiority. In this sweeping humiliation of all past historical authorities, "the discovery of the historical character of the biblical writings" is the point of departure.[14] All things are measured by the "standards which the autonomous mind had set up for its own purposes." Critical knowledge "arrogates to itself the right and the power to decide the truth and falsehood of the Word of God."[15]

Orthodox dogma in this historicizing schema is simply a prejudice, especially the doctrine of sin, which is couched in terms of slavery. It becomes most hated, because it soundly denies objectivity in divine matters. "One must be unconditionally free from all preconceived opinions and all spiritual prejudice" to be a modern critic.[16] This aspect of modernism is not a technical dogmatic error about the nature of Scripture. The idea of God telling man anything becomes irrelevant, as meaningless as any other old author who cannot speak to future "moderns." The attack on inspiration, in many and various ways since the Enlightenment, expresses this new approach to knowledge: facts are established by observation and compelling, rational cogency.

Revelation is Only Historical

The scientific approach to history rules religion. Revelation is implicitly denied because "revelation cannot be historically verified," therefore it is deemed uncertain.[17] There is left only natural, verifiable religion, in which reason is the primary authority. A non-revelatory Christianity is

---

[13] G. Ebeling, *Word and Faith*, 47.
[14] K. Scholder, *Birth of Modern Critical Theology*, 125.
[15] Erwin Reisner, quoted in: G. Ebeling, *Word and Faith*, 18–19.
[16] K. Scholder, *Birth of Modern Critical Theology*, 136.
[17] Carl E. Braaten, *History and Hermeneutics* (Philadelphia: Westminster Press, 1966), 25.

essentially deistic in orientation. Deism did not so much become a new religion in Enlightenment thinking as Christianity became deistic.

To a modern person, a miracle is "a transgression of a law of nature by a particular volition of the Deity, or by the interposition of some invisible agent."[18] While he thinks of miracles as those special events of Christianity most attacked, it was rather the naturalistic idea of "miracle" that was introduced by rationalism. What is not in the special category of "miracles" is explainable without divine interaction. The Bible, however, knows no word corresponding to "miracle," as an exceptional intervention of a mostly absent God. In Him, who upholds the universe by the Word of His power, we live and move and have our being (Heb. 1:3; Acts 17:28). No biblical fact can be stated without referencing God's Spirit. Nothing exists apart from the living God. Everything is therefore religious or spiritual.

If there is no connection between God and this world, man is unlimited in his field of observation. Then the scientific paradigm for knowledge can reign supreme. Natural laws get the credit for this modern world, so even God has nothing to do with it presently. Even if God is intellectually declared behind the laws, it is still "a demystification of the world."[19] "No theoretical result of that exploration of the world which men call science has been more fruitful than the conception of natural law." But it was "fiercely resisted by theologians from the beginning as something undesirable and subversive."[20] It allowed God to be criticized objectively as an idea, because He was declared practically unnecessary. The splitting of God from nature, previously understood as general revelation, led next to the humanizing of the Bible—His special revelation. "The historical approach to Biblical literature is one of the great events in the history of Christianity and even of religion and human culture."[21]

Whereas wonders, prophecies, and signs recorded in Scripture set it apart in the pre-modern era, they seemed to discredit it to modern critics.

---

[18] David Hume, "Of Miracles" (1751), quoted in: *Biblical Errancy*, ed. N. L. Geisler, 31.

[19] Peter Harrison, *The Bible, Protestantism, and the Rise of Natural Science* (Cambridge: Cambridge University Press, 2001), 7.

[20] J. Y. Simpson, *Landmarks in the Struggle between Science and Religion*, 144.

[21] Paul Tillich, quoted in: C. E. Braaten, *History and Hermeneutics*, 33.

> When the Church proposes the doctrine of Divine Revelation and of holy Scripture to people for the first time, what signs does she offer that it is really the Word of God? Signs of this are the following: 1) The sublimity of this doctrine, which shows that it can not be any invention of man's reason. 2) The purity of doctrine, which shows that it is from the all-pure mind of God. 3) Prophecies 4) Miracles 5) The mighty effect of this doctrine upon the hearts of men, beyond all but divine power.[22]

The Bible's own internal criteria fell into suspicion, following the pattern of "natural laws." But it is revelation which certifies all true religion, including Christ's life. "The rejection of revelation is the modern apostasy."[23] Revelation is a fascinating hypothesis to moderns, but never is it equated with Scripture itself. Nothing could verify revelation itself or make it believable in the face of modern rationalism. When man is seen as the ground of certainty, a plain, literal revelation is a self-contradiction—a denial of man himself.

The "complete" scientific descriptions of science replaced the two books analogy of traditional Christianity: "The whole sensible world is a kind of book written by the finger of God."[24] Not just the book of Scripture, but also the book of general revelation, the creation itself, speaks of God's character directly: "For his invisible attributes, namely, his eternal power and divine nature, have been clearly perceived, ever since the creation of the world, in the things that have been made" (Rom. 1:20). History in the modern sense, though, can speak only with human authority. "If the theologian believes that the events of the Bible are results of the supernatural intervention of God, the historian regards such an explanation as a hindrance to true historical understanding."[25] Scientific assumptions, rather than the facts themselves, make history atheistic.

The modern idea of "miracle" is a deistic assumption that God is not normally involved in His creation. "For ancient peoples miracles

---

[22] *The Longer Catechism of the Orthodox, Catholic, Eastern Church* (Moscow, 1839). P. Schaff, *Creeds of Christendom*, 2:454.

[23] H. D. McDonald, *Theories of Revelation*, 1:134.

[24] Hugh of St. Victor (d. 1141), quoted in: P. Harrison, *The Bible, Protestantism, and the Rise of Natural Science*, 1.

[25] Van A. Harvey, *The Historian and the Believer: The Morality of Historical Knowledge and Christian Belief* (Philadelphia: Westminster Press, 1966), 5.

appeared to be common occurrences, because these people did not know about the existence of the laws of nature."[26] Now, natural laws receive all credit and reverence, as if nature created and maintains itself. The Bible states that Christ directly "upholds the universe by the word of his power" (Heb. 1:3). "This nature is the great, incomprehensible miracle. We live in the miracle."[27] It is highly instructive that "the Hebrew people lacked a word for 'nature'."[28] This distancing of God from the created world ushered in further theological attempts to prevent His speaking to man in Scripture. The Cartesian method rules out what it cannot apprehend.

It is unsurprising that once revealed knowledge was denied, natural knowledge would be also. The two are connected by virtue of their divine Author. The clearest and highest revelation is in written form, not in His general works of power now in bondage to decay. Nature in the old dogmatics is the general revelation of God, since it speaks of the nature of God. The modern view of nature, based on man, changed the view of Scripture. Samuel Taylor Coleridge (d. 1834) was "acutely aware of the need for a reformulation of the traditional doctrine of Scripture's revelation in light of the new historical knowledge."[29] The modernist urge is to deny the authority of religion, while attempting to salvage its contents. The doctrine of inspiration has borne the brunt of this reorienting of religious knowledge.

The conflict of scientific methods and religious claims is unavoidable. The essence of historical criticism is "the uncompromising supremacy of 'scientific' human reason."[30] Inspiration as a supernatural ground for facts became a stumbling block in the scientific approach to history. The freedom to criticize has become so ingrained that to submit to any authority besides reason becomes "rationalistic and legalistic." A basic assumption of modernity is that man is not "*capax verbi Domini,*" that is, capable of receiving the eternal Word of God.[31]

Without revelation Christianity is a natural religion. In a strange

---

[26]Gary R. Habermas in *Biblical Errancy*, ed. N. L. Geisler, 33.

[27]F. Bettex, *The Bible the Word of God*, 188.

[28]H. D. McDonald, *Theories of Revelation*, 1:87.

[29]A. Richardson, *The Bible in the Age of Science*, 57.

[30]Kurt E. Marquart, *Anatomy of an Explosion: Missouri in Lutheran Perspective* (Fort Wayne, IN: Concordia Theological Seminary Press, 1977), 114.

[31]H. D. McDonald, *Theories of Revelation*, 1:151.

turnabout, historical critics redefine revelation as historical. Not words but events came under the auspices of the critic. Truth as event allows all so-called "truths" to be compatible with the "real" facts established scientifically. "In this event God encounters human beings through his Word, while at the same time remaining completely free in his decision to (or not to) reveal himself."[32] While recasting Christ's incarnation and death as revelatory seems quite promising, modern history is still man-centered. In allowing no contradiction with the facts of science, it permits no truth according to any meaningful definition.

Inspiration as indirect influence of the Spirit is acceptable to moderns—as long as history remains subjugated: pre-modern dogmaticians "misuse the doctrine of inspiration to transcend the historical relativity of the Scriptures."[33] If Scripture is wholly a divine product, it is not limited to a certain audience, time period, or culture. Here is the fountain of modern uneasiness with inerrancy: inerrancy is incompatible with historical distancing, since inerrancy across time is not a property of man's limited authority.[34] The bedrock assumption of modernism is that truth, in so far as it applies to people, is relative and not over the modern man. "It is not possible to be historical and unconditioned at the same time."[35] It is not the doctrine of Scripture that is directly attacked: everything is made relative—Christian doctrine and the biblical facts are distanced and made inconsequential. In this mindset "supposedly absolute 'dogma' . . . [are] actually matters of historical contingency," and the Bible is a mere historical artifact of an unenlightened people.[36]

"Truth as being historically-relative," means inerrancy, in the sense of timeless authority, does not compute.[37] Revelation, instead of being a divine communication to man, is put into the critical closet of history, where God cannot speak absolutely. "Revelation is not so much . . . the disclosure and communication of general timeless truths," but "primarily

---

[32] P. H. Nafzger, *These Are Written*, 44.

[33] Warren Quanbeck in *Studies in Lutheran Hermeneutics*, ed. John Reumann (Philadelphia: Fortress Press, 1979), 180.

[34] H. Lee, "Accommodation: Orthodox, Socinian, and Contemporary," 340.

[35] Reinhold Niebuhr, quoted in: Kenneth S. Kantzer, "Revelation and Inspiration in Neo-Orthodox Theology, Part II: The Method of Revelation," *Bibliotheca Sacra* 115.459 (1958), 222.

[36] Alister E. McGrath in *The Science of Theology*, vol. 1 of *The History of Christian Theology*, ed. Paul Avis (Grand Rapids, Eerdmans, 1986), 221.

[37] H. Lee, "Accommodation: Orthodox, Socinian, and Contemporary," 345.

and properly a definite event . . . the appearance of Jesus Christ."[38] Truth is redefined in terms of action, as typified in the dynamic-only definition of inspiration. "Revelation is something that happens," not the communication of information.[39]

The modern premise is that all knowledge is historically subjective, including revelation itself. Past deeds cannot touch present man. So revelation is described as a personal encounter, which cannot be scientifically errant or false. This is all too convenient, because a personal, non-verbalizing authority cannot command or critique man's actions and beliefs. Man wants the right to criticize everything, so God is revised. For the modern, He can only "reveal" Himself impersonally and wordlessly. God is merely an idea without authority. The sock of philosophical naturalism will not permit Him to vocalize a single word.

Historical events are critical to Christianity. But witnessing Jesus' execution did not give faith. Touching Jesus did not heal or save without Spirit-wrought faith. It did not look as if God was dying for the sins of the world on the cross. The bare event is not helpful or vivifying, without a word of explanation. As important as historical truthfulness is, without an authoritative word concerning events, it is not something in which to believe. Man cannot access past history.

All action is God's. History is "His Story," but natural, historical knowledge never attains to the Gospel. The modernist–fundamentalist divide can be seen very clearly in the definition of revelation: it is either only action or "always communication."[40] There is no middle ground in this fundamental supposition.

No current revelation is allowed in modernist thought. Scripture, instead, bears witness to past revelation and makes a present revelation possible.[41] Yet history, which is always time-conditioned, is incidental to modern observable truth. Gotthold Ephraim Lessing's (d. 1781) infamous dictum speaks of the "accidental truths of history," noting

---

[38] G. Ebeling, *Word and Faith*, 29.

[39] Emil Brunner, quoted in: K. Kantzer, "Revelation and Inspiration in Neo-Orthodox Theology," 219.

[40] Bernard Ramm, *The Witness of the Spirit: An Essay on the Contemporary Relevance of the Internal Witness of the Holy Spirit* (Grand Rapids: Eerdmans, 1959), 30.

[41] P. H. Nafzger, *These Are Written*, 41.

that historical facts have no bearing on rationally established truth.[42] As a modernist LCMS professor puts it:

> No text, biblical or otherwise, can prove the historicity of any event. Moreover, no analysis of literary aspects, of the genre, or of the function of a text can prove or disprove the historicity of any event described in the text. Events and texts belong to two phenomenologically distinct categories that do not intersect. As a result, the most that we can learn from a text is what the author believed to have happened, or what he wants the reader to believe about what has happened.[43]

In the elevation of reason, all history, including that of Jesus, was dealt a fatal blow. "For modern man everything, the whole of reality, turns to history" and "what is not historical is for modern man not real."[44] History is critically insulated from moderns—it has no authority. Modernism relativizes everything by submitting all things to man's objective evaluation. Divine words are thought impossible: "God, in his divine freedom, chooses to exalt human speech and speak about himself through the words of his commissioned spokesman."[45]

According to God Himself, He is involved in all history. From a bird that falls, to men who live and move and have their being in Him, to pagan leaders, nothing happens apart from God the Creator (Mt. 10:29; 2 Chron. 36:22).[46] However, God's imprecise (to us) involvement with all things is not salvation. Only a communication can explain historical actions and make them known to us. Faith exists on account of God's promises which reveal His hidden actions, not bare events.

If revelation is only historical and history itself is made trivial and distant, what is left of Christianity? In that case, Scripture can only be an aid to natural theology, not the intrinsic theological authority.[47] The Bible is declared by fiat to be patently insufficient in the wake of

---

[42] *Challenges to Inerrancy,* eds. G. R. Lewis and B. A. Demarest, 27.

[43] David L. Adams, "Some Observations on the Historicity of the Biblical Creation Account," in *Listening to the Word of God: Exegetical Approaches,* eds. Achim Behrens and Jorg Christian Salzmann (Göttingen: Edition Ruprecht, 2016), 13–14.

[44] G. Ebeling, *Word and Faith,* 363.

[45] P. H. Nafzger, *These Are Written,* 41.

[46] God moved Cyrus' spirit to cause him to speak and write, but He did not speak directly through him as with the prophets. This was done "that the word of the Lord by the mouth of Jeremiah might be fulfilled."

[47] *Challenges to Inerrancy,* eds. G. R. Lewis and B. A. Demarest, 24.

reason's new-found sufficiency. It is an axiom of biblical criticism that "all revelation is submerged in a swap of the history of ideas, the history of religion, and the history of culture."[48] Scripture and all the truths of Christianity are de-divinized and made thoroughly human. In this method all things become subject to twisted, rebellious man.

Karl Barth (d. 1968) is a recent peddler of history as revelation, but it is much older. A follower of Coleridge in the early 1800s held to the "vehement rejection of the idea of propositional revelation, so that revelation is illumination, not communication."[49] This becomes the standard modern play to neuter Scripture and still talk of Christian truth. So historical criticism does not deny the supernatural or miracles: it sets up man as the center of truth in direct contradiction to God Himself. If God's Word is historically conditioned, there is no actual eternal Word of God: "dynamic inspiration rests on God's decision to speak his Word ever and again in the history of the Church and throughout the text of the Bible."[50]

The Scriptures are not seen as the absolute authority today, so Christianity is completely altered. "The total phenomenon of Christianity was now seen in a historical context," making it completely foreign and alien to the modern observer.[51] The severity of this approach is hard to see for us because we have been steeped in it. We are trained to be critical towards everything, and only make exceptions with great difficulty. Everything, even inspired words, are assumed to be in a historical cause and effect circle. "It is only when we regard [the Scriptures] as historical evidence that their true relevance appears."[52] To be a modern is to share this commitment to divining truth through reason and experience. "The modern, educated, liberated Christian can put [dogmatic opinions] in appropriate historical niches."[53] Nothing touches him.

"Historical criticism is marked by a commitment to the sovereignty

---

[48] Olav Valen-Sendstad, *The Word That Can Never Die: A Scriptural Critique of Theological Trends* (St. Louis: Concordia Publishing House, 1966), 25.

[49] H. D. McDonald, *Theories of Revelation*, 1:190.

[50] S. J. Nichols and E. T. Brandt, *Ancient Word, Changing Worlds*, 37.

[51] Allan D. Galloway in *The Science of Theology*, ed. P. Avis, 290.

[52] A Concordia Seminary, St. Louis professor (1970–71), quoted in: P. A. Zimmerman, *A Seminary in Crisis*, 174.

[53] K. Hagen, *Foundations of Theology in the Continental Reformation*, 227.

of human reason."[54] The "critical" aspect of historical criticism is not found in its conclusions, but the elevation of the critical interpreter who approaches all of history as judge and jury. It rests on the fundamental presupposition that man is lord over all history, and therefore all sources of history, including Scripture. Anything above him is ruled out by principle, unless man after consideration allows such authority as an intellectual decision. History "expresses the modern view of truth—a view which compels us where history is concerned to make a critical distinction between the tradition of history and the facts of history, between the picture that has been handed down of an event and the reality of the event itself."[55] Only by this method, centered squarely on man, is assured, truthful knowledge found. Man, by following this scientific method, establishes true reality—since all things have a historical character.

In contrast to the assumed divine authority of man, "the God of most contemporary theology is a sort of deaf-mute who can act [historically] but not speak."[56] Not individual teachings but the very "absoluteness of Christianity was under the severest attack."[57] This was a "revolution of theological thinking more profound than any that had taken place in any previous century of the Church's history, not including the 16th." Even that qualification is doubtful, since historical science allows an "undogmatic study of the Bible and Christian doctrine" by the very teachers of Christianity.[58]

---

[54] *Studies in Lutheran Hermeneutics*, ed. J. Reumann, 43.
[55] G. Ebeling, *Word and Faith*, 290.
[56] K. Kantzer, "Revelation and Inspiration in Neo-Orthodox Theology," 225.
[57] Allan D. Galloway in *The Science of Theology*, ed. P. Avis, 292.
[58] A. Richardson, *The Bible in the Age of Science*, 62.

Chapter 11

Scientific "Facts"

The idea that facts are only established by the scientific method came to prevail in the modern age. "One of the most distinctive features of the emergence of a scientific culture in modern Europe is the gradual assimilation of all cognitive values to scientific ones."[1] Factual data in this scheme are only naturally known and observable by man, who chooses to verify them. Carried to the extreme, a double truth results. "Theological and religious inerrancy" is not allowed to conflict with truth, that is, "the rigid theory of factual inerrancy." The highest truth is clearly with evidential rationalism, since error is distinctly possible with scientific depictions. So the "lower truth" of religion must submit to the higher truth of science and its authority—the individual. Science "is a new sort of religion of Nature, which has entirely displaced Christianity from the thought of a large part of our generation."[2]

Instead of completely rejecting Christianity, which would be the consistent choice, Christianity was "updated" or modernized. The factual parts that troubled "modern man" were declared nonessential. "Historical criticism evolved from the seedbed of modernism, a cultural mood that gave prominence to hard sciences and stressed rationalism."[3] That priority is not often stated, but critics do have an authority they bow to: the idol of the self and the intellect. The very word "science"

---

[1] Stephen Gaukroger, *The Emergence of a Scientific Culture: Science and the Shaping of Modernity 1210–1685* (Oxford: Clarendon Press, 2006), 12.

[2] H. D. McDonald, *Theories of Revelation*, 2:210, 2:21.

[3] S. J. Nichols and E. T. Brandt, *Ancient Word, Changing Worlds*, 112.

is worshiped as all-authoritative. The unscientific modern man bows to "truckloads of science," as if it were a substance that offered a final conclusion, instead of a negative method centered on doubt.

The clearest modern theologians are honest: "Evolution . . . has become the scientific truth, with which all honest theology has to come to terms."[4] This assumption is simply bowing down to an alien, non-theological truth: "Part of the intellectual honesty of adult [enlightened] man is that in the area of faith he will accept no truth-claim that conflicts with scientific knowledge."[5] Moderns readily believe in inerrancy within the scientific realm, but when statements do not deal with hard facts, inerrancy is inappropriate.

In the battle of the relationship of the sun to earth, the issue was the inspiration, and therefore the truth, of Scripture. The Lutheran Johann Kepler, an astute theologian and scientist, spoke of the great fear that the Holy Spirit (in Scripture) would "be burdened with a lie."[6] While Creation is a "stumbling-block to all educated persons" today, that was not the case for the first generation of physical scientists.[7] "We often describe how things appear to us, even if we know that they are in fact different," Kepler notes.[8] It is not the Bible that conflicts with extra-scriptural truth: "The chief impediments to improvements in natural philosophy lay not with the text of Scripture, but with the decrees of the Holy [Roman] Office and the opinions of the Doctors of the Church."[9] Reason does have authority over creation, but its examination will never be complete or very full without divine knowledge of the Creator. Science as ultimate truth is truly a religious claim.

Modern theology is concerned with bringing "theological statements into harmony with the [new] sense of truth as this had been transformed by historical and scientific study."[10] The atheistic method did not discover truth; instead, it reshaped truth to be compatible with the

---

[4] E. Brunner, *Revelation and Reason*, 279.

[5] Helmut Thielicke, *The Evangelical Faith: Prolegomena: The Relation of Theology to Modern Thought Forms*, vol. 1 of 3, trans. and ed. Geoffrey W. Bromiley (Edinburgh: T&T Clark, 1978), 66.

[6] K. Scholder, *Birth of Modern Critical Theology*, 55.

[7] F. Bettex, *The Bible the Word of God*, 178.

[8] K. Scholder, *Birth of Modern Critical Theology*, 55.

[9] P. Harrison, *The Bible, Protestantism, and the Rise of Natural Science*, 113.

[10] H. Thielicke, *Prolegomena*, 32.

facts of its new "faith." The salvation it offered was purely materialistic, reinforcing its prominent status. The reverence for science involves "an implicit faith that by the methods of physical science, and these methods alone, could be solved all the problems arising out of the relation of man to man and man towards universe."[11]

Biblical inerrancy, if seen in the scientific light, can only play into the hands of its deniers. The scientific attitude, in opposition to the doctrinal mindset, made inerrancy problematic and impossibly burdensome. The Christian would have to verify and prove rationally every claim before critically asserting Scripture's inerrancy. The direct inspiration of Scripture, which necessarily means its inerrancy, is seen by moderns as the "greatest tragedy in Church history." The orthodox doctrine of verbal inspiration "was a catastrophe for the whole fabric of the doctrine of the Church."[12] This was held because just one minor error could topple the Gospel of Christ. Again, it is not specific scientific conclusions which cause this opinion, but scientific principles universally applied.

It was not the actual discoveries or the scientists themselves which contradicted Scripture, but ecclesiastical overreactions and materialist philosophies. After all, we do not say those who speak of the "sunrise" and "sunset" are erring and untruthful, even today. The issue is not reasonable scientific facts but the spiritual priority of natural facts over revealed facts which became the crux of modernism. "The scientific and historical thinking of the modern age has destroyed the naïve attitude of accepting without question that everything the *Bible* says is fact."[13] However, the Christian who accepts Scripture need have no fear that the facts of the world which Christ made will contradict what He speaks. To assume otherwise is to invent a different Christ from the scriptural One.

The new method of determining truth sought visible evidence. Proof, which used to imply citing an authority, must now be rationally demonstrable. The result of Scripture's inspired words was the opposite of modern science. The "rigid adherence to the letter of Scripture [had]

---

[11] Beatrice Webb (d. 1943), quoted in: S. Gaukroger, *The Emergence of a Scientific Culture*, 24.
[12] E. Brunner, *Revelation and Reason*, 37, 274.
[13] G. Ebeling, *Word and Faith*, 195.

proved so disastrous."[14] Confessional adherence by intractable parties led to visible atrocities, such as the Thirty Years War (1618–1648). This new standard of truth is results-oriented. Proof was not in external authority, but in evidence clearly observable by all: the advances of modern culture.

"We have become so tolerant of religious pluralism that the last thing we want to do is to argue, let alone fight, about religious differences."[15] But the incarnate Lord did speak and make His will known. He spoke "truly, I say to you" with the greatest authority, and His words still carry the exact same weight. It is not just His historical words which should be red-lettered, but the entire Scripture. Since the Spirit and Christ are one with the Father, there can be no separation of inspired communication from Christ.

A claim to divine teaching does divide—those who do not accept it are accused of disobeying God. The literal fundamentalist sees division as divinely warranted and also warranting his position: "for there must be factions among you in order that those who are genuine among you may be recognized" (1 Cor. 11:19). The modernist, however, overrules all supernatural claims because he has the "proof" of peace in the universal atheism which allows no authoritative convictions. The fact that "fundamentalists" look forward to heaven and do not take consolation in a technologically improved earth is ridiculous to moderns. The two factions operate with vastly different methods for the verification of facts. One seeks present results, while the other seeks heavenly vindication and recompense for the repercussions suffered temporally on behalf of the truth.

The modern seeks only scientific truth. "The real Christ is only accessible through historical-critical analysis."[16] What does Scripture say after being fed through the meat grinder of historical science? It tells of a deistic, man-centered, natural religion which can never free people from their sins. All it is left with is the "awkward notion of the deity spending eternity in solitude and idleness," having no real

---

[14] J. Y. Simpson, *Landmarks in the Struggle between Science and Religion*, 137.

[15] K. Hagen, *Foundations of Theology in the Continental Reformation*, 226–227.

[16] Stanley J. Grenz and Roger E. Olson, *20th-Century Theology: God & the World in a Transitional Age* (Downers Grove, IL: InterVarsity Press, 2010), 270.

interaction with man.[17] "No facts, no good news; no good news, no hope. The Bible is quite useless unless it is a record of facts."[18]

The battle between the Bible and science is for preeminent authority and status, not the specific truth of individual facts. Moderns today accept evolution on flimsy evidence without any critical reservation. Francis Bacon (d. 1626) rallied the cry for the complete separation of "the realm of reason and science from the realm of faith and religion."[19] It led to "the emancipation of science (and the scientist), who regardless of the doctors of the church feels bound only by truth which can be objectively demonstrated and proved."[20] This grave distinction dissolves timeless authority, and therefore all of Christianity.

"The watershed between the Middle Ages and modernity" is that "experience and proof are set over against the traditions," even Scripture.[21] The uniqueness of Scripture, specifically its divine origin explained by inspiration, became most problematic. The upshot of the naturalistic conception of truth is that man is not limited by God in his investigations. At the start, by the divine right of method, man is preeminent over all things. This temporary assumption in scientific judgment has become a cultural disease. But man can only get authority by distancing himself from God—declaring himself free. Atheism *is* powerful—powerfully damning. It allows man to judge everything, even God, so he may think as a god.

In critical modernism, no true revelatory inspiration is allowed. The writers, ideas, thoughts, and readers may be "inspired," but the Bible we actually possess is just human words. Since a good portion of the Bible is historical in character, this shift in interpretation changed Christianity. Christianity is concerned with historical acts, especially those done by Jesus. The traditional view is that, "God's truth can never be at variance with the phenomena of God's world."[22] The same

---

[17] E. L. Long, *Religious Beliefs of American Scientists*, 42.

[18] J. Gresham Machen, quoted in: S. J. Nichols and E. T. Brandt, *Ancient Word, Changing Worlds*, 123.

[19] *Inerrancy*, ed. N. L. Geisler, 313.

[20] K. Scholder, *Birth of Modern Critical Theology*, 58.

[21] K. Scholder, *Birth of Modern Critical Theology*, 60–61.

[22] J. L. Porter, "Science and Revelation: Their Distinctive Provinces," in *Science and Revelation: A Series of Lectures in Reply to the Theories of Tyndall, Huxley, Darwin, Spencer, etc.* (New York: Scribner, Welford & Armstrong, 1875), 35.

Scientific History: The Subjugation of History

God who speaks in Scripture, made the world—both are His books. The view of the Bible until the late 18th century was that it purported "to describe what actually happened in the world of space and time."[23]

Authoritative criticism necessitates a "historical distance between the word of Scripture and its contemporary reader."[24] God is actually pushed away, so He may be studied objectively. This has become the "neutral" method of sinful man. It was not so much a theological attack, but the "claim of modern science to infallibility . . . [which] essentially helped to shape the modern mind."[25] Scientific facts did not undo orthodoxy; rather, it was the unscientific theologians whose minds were shaped by scientific values. Without a greater, revealed truth, the scientific approach redefines truth itself, along with the entire Christian religion. "It was primarily biblical criticism and history rather than science that were the external causes of the intellectual rethinking of religious sensibilities and sources of authority."[26] Scripture, tradition, and personal revelations were all swept away by the new scientific approach to religion.

This destructive criticism has not stopped moderns from using and perverting the Bible. "Like all ancient literature the Old and New Testaments know nothing of the distinction of fact and value which is so important to us, between history, on the one hand, and saga and legend on the other."[27] Theology, when it gives up factuality, which it has in the main, becomes the attempt to give meaning and significance without real, concrete truth. "Till now—at least in the last two centuries—the truth of revelation has been subordinated to the judgment of historical science."[28]

---

[23] James C. Livingston, *Modern Christian Thought, Vol. 1: The Enlightenment and the Nineteenth Century* (Upper Saddle Creek, NJ: Prentice Hall; 2nd ed., 1997), 238.

[24] Karl Girgensohn, quoted in: Gerhard Maier, *Biblical Hermeneutics,* trans. Robert Yarbrough (Wheaton, IL: Crossway, 1994), 259.

[25] K. Scholder, *Birth of Modern Critical Theology,* 171.

[26] S. Gaukroger, *The Emergence of a Scientific Culture,* 23–24.

[27] Karl Barth, quoted in: *Studies in Lutheran Hermeneutics,* ed. J. Reumann, 110.

[28] Erwin Reisner (1947), quoted in: G. Ebeling, *Word and Faith,* 18.

Chapter 12

Methodological Atheism

Criticism is an approach rooted in the doubt of external authorities. The assumption is that truth is rational, so man may understand it. "Science promoted an intellectual arrogance which led man to prefer his own notions to the inspired Word of God."[1] Thus historical criticism as a method describes a basic attitude towards Scripture. Since the Word is the only means of knowing God, the way in which the Word of God is approached is simultaneously the attitude toward God Himself. "Modern science follows the principle of Cartesian doubt, accepting nothing as certain that admits of any doubt whatsoever."[2] No method could be more incompatible with Scripture, which is given to dispel doubt and give comfort in Christ. "For whatever was written in former days was written for our instruction, that through endurance and through the encouragement of the Scriptures we might have hope" (Rom. 15:4).

The rigorous application of Cartesian doubt is a universal method of "neutral" destruction, equalizing all traditions and witnesses. This "scientific atheism" is a suspension of judgment, an initial agnosticism, so that man may judge all things.[3] Consistent rational criticism will doubt all supposed authorities, except the internal, rational one—at least initially. "It is characteristic of its procedure that the scientist in

---

[1] R. S. Westfall, *Science and Religion in Seventeenth-Century England*, 24.
[2] Hans-Georg Gadamer, quoted in: G. Maier, *Biblical Hermeneutics*, 23–24.
[3] J. A. Zahm, *Evolution and Dogma* (Chicago: D. H. McBride, 1896; reprint, Hicksville, NY: Regina Press, 1975), 254.

asking his questions disengages himself from the object of his research."[4]

Even belief "in God" has become largely atheistic, because it refers to the *scientific* fact of His existence, not any confessional stance about His will. The word "god" is generic and deistic. This common phrase, "I believe in [a] god" is a product of minds which see truth as scientific in character. It supposes that the hearer of the phrase should care that man validates God's existence. It is completely opposed to Christian belief which confesses specific divine teachings which God Himself has revealed.

Modern "belief" has become thoroughly atheistic, a critical, man-centered verification of God's hypothetical existence—an authorization, not a profession. "The Enlightenment devotion to reason [is] the modern form of idolatry."[5] The focus on the individual as the judge and focal point of truth has altered society much more than technological innovations. "Pre-Cartesian atheism was of limited appeal," but "modern atheism is militantly universal in its appeal" as an indispensable methodological assumption.[6]

In traditional thinking, atheism was not only a factual denial of God's existence, but a way of thinking or living which distanced God from man.

> By Atheism we mean any system of opinion which leads men either to doubt or to deny the Existence, Providence, and Government of a living, personal, and holy God, as the Creator and Lord of the world. In its practical aspect, it is that state of mind which leads them to forget, disown, or disobey Him.[7]

Cartesian methodology involves a universal prejudgment, a fundamental attitude of unbelief toward all truth claims. Historical criticism is not a doctrinal error in itself, but it allows all doctrines to be weighed by human reason and judged by arbitrary standards. God is not like a grasshopper or leaf that man can study. The method of science which

---

[4] Gustaf Wingren, *Theology in Conflict: Nygren, Barth, Bultmann*, trans. Eric Wahlstrom (Philadelphia: Muhlenberg Press, 1958), 21.

[5] J. C. Livingston, *Modern Christian Thought*, 71.

[6] Cornelio Fabro, *God in Exile: Modern Atheism: A Study of the Internal Dynamics of Modern Atheism, from its Roots in the Cartesian Cogito to the Present*, trans. Arthur Gibson (Mahwah, NJ: Paulist Press, 1968), xv–xvi.

[7] James Buchanan, *Modern Atheism: Under its Forms of Pantheism, Materialism, Secularism, Development, and Natural Laws* (Boston: Gould and Lincoln, 1857), 15.

works reasonably well for created objects does not work for the Creator at all. Revelation can only be verified by God Himself.

> Historical criticism evaluates every human report about the past—even historical remembrance—as a source limited and at least partially biased due to human subjectivity. Modern critical historiography wants to determine how heavily a historical report is influenced and eventually distorted by an observer's or author's philosophical presuppositions and prejudices.[8]

It is unscientific to make an exception for Scripture beforehand. Because doubting God's Word is doubting God, "one can no more be a little historical-critical then a little pregnant."[9]

Historical criticism is confessed to be a "neutral methodology" because man may possibly verify the Scriptures as true, if its evidences are strong enough. But this approach to truth rules out divinely revealed facts. Historical criticism treats the scriptural books as disparate witnesses in the critical court of law to "be interrogated and their answers evaluated," just like all other witnesses.[10] In this approach the Bible needs man to bestow divine status on God's words. Therefore historical criticism is hardly neutral for every topic, unless one were actually the highest authority, that is, a true god. To the consistent critic "idols are the prejudices or limitations to which the mind is subject."[11] That means that divine truth, literally given, is also an idol to moderns, because it limits and restricts reason.

Man in critical methodology is the highest authority. His tool is universal doubt. "The reflecting I is brought to the centre of the universe," so that "God is the I of [Descartes'] self-awareness."[12] This is a basic "prescientific decision to be 'as God'."[13] This is rightly seen as empowering and revolutionary, so that the historian, the independent "thinker," is "radically autonomous."[14] The critic, that is everyone, assumes the right

---

[8] D. M. Farkasfalvy, *Inspiration and Interpretation*, 122.
[9] Eta Linnemann, *Historical Criticism of the Bible: Methodology or Ideology?* (Grand Rapids: Baker Books, 1990), 123.
[10] E. Krentz, *The Historical-Critical Method*, 42.
[11] This was "one of [Francis] Bacon's [d. 1626] favorite themes." C. Hill, *Antichrist in Seventeenth-Century England*, 45.
[12] K. Scholder, *Birth of Modern Critical Theology*, 111.
[13] Erwin Reisner, quoted in: G. Ebeling, *Word and Faith*, 19.
[14] V. A. Harvey, *The Historian and the Believer*, 42.

to judge all witnesses. Man is supreme—this is the neutral starting point of the scientific method: "critical human intelligence as arbiter of truth."[15] This is accurately called "hermeneutical atheism."[16]

Inerrancy, as a property of truth, became associated with rational, empirical knowledge—not religion. The scientific method guarantees factual truth, much as inspiration meant biblical infallibility. This Cartesian method of doubt has led to incalculable damage. Inspiration of the divinity is moved from outside of man and centered within him. "The consciousness of the infallibility of their method which is expressed to a greater or lesser degree by all the Cartesians is a characteristic of the new spirit."[17] The new "spirit" is aptly called "modern." Inerrancy is not denied in general; it is rather associated with the internal mental machinations of man, not any "sacrifice of the intellect" required by authoritative revelation.

In reality, it is not pure critical thinking based on the evidence that leads people to use historical criticism, but simple peer pressure in the academy. Professional physical scientists are not more atheistic than the general population. While the professional theologian likely knows little of true science, he knows the desirable prestige and popular authority of science. The modern has removed "God far from reality to comply with so-called scientific demands of a modern, secular worldview which allows no room for divine intervention." Even if God is allowed to "sin" against natural laws, He is not allowed to work in His creation without very convincing testimony. The starting assumption is not that "all things were created through [Christ] and for him . . . and in him all things hold together" (Col. 1:16–17). The starting point is instead "factual atheism."[18]

It reminds one of the saying that "to one with a hammer, everything is a nail." While a hammer may be a "neutral methodological tool," it is not appropriate for every surface. "But the Reformers, like the

---

[15]Kurt Marquart in *Studies in Lutheran Hermeneutics*, ed. J. Reumann, 318.

[16]Graeme Goldsworthy, *Gospel-Centered Hermeneutics: Foundations and Principles of Evangelical Biblical Interpretation* (Downers Grove, IL: InterVarsity Press, 2006), 53.

[17]K. Scholder, *Birth of Modern Critical Theology*, 115.

[18]Klaus Bockmühl, *The Unreal God of Modern Theology: Bultmann, Barth, and the Theology of Atheism: A Call to Recovering the Truth of God's Reality* (Colorado Springs, CO: Helmers & Howard, 1988), 2.

church leaders and Christian interpreters before them, would not have countenanced the priority of method over theological instruction, or the possibility of reading and teaching largely without oversight."[19] Secular understandings have taken the place of the actual teaching of God's Word. "The rise of Biblical scholarship made necessary a new doctrine of the inspiration of Holy Scripture."[20] Method is the new sacred doctrine, passed from teacher to student with the gravest reverence.

"In 1600, western civilization found its focus in the Christian religion; by 1700, modern natural science had displaced religion from its central position."[21] Historical criticism has become institutionalized heresy. It does not allow any doctrine to be unfettered by human judgment. A leading scholar divulged: "we can no longer think without or against this method."[22] This unhealthy devotion to method intoxicates with the supposed power of reason. It quickly becomes idolatry, to which the civilized scholar must bow. If one "wishes to avoid the charge of Biblicism and Fundamentalism in scholarly circles, one must find some way to square the acceptance of historical-critical scholarship" with the plain words of Scripture, especially those passages which conflict with accepted scientific ideas.[23]

Appealing to the Spirit is unscholarly and absurd. He is non-rational and super-rational, since God is above man. As a consequence, the doctrine of the internal testimony of the Spirit is considered obsolete by critical practitioners. In pre-modern explanations, it affirms that the Spirit Himself, not human criteria, convinces one of the authority of His Word. "Claiming the *testimonium* [testimony of the Spirit] as proof of inspiration is not without problems. It comes close to a tautology, since accepting the Spirit's testimony is faith. It suspiciously resembles Calvin's view of the indwelling of the Holy Spirit as evidence of faith."[24] This is simply an uncritical acceptance of harmful philosophy, that one

---

[19] Daniel J. Treier, "The Superiority of Pre-critical exegesis?: Sic et Non," *Trinity Journal* 24:1 (2003), 101.

[20] H. D. McDonald, *Theories of Revelation*, 2:219.

[21] R. S. Westfall, *Science and Religion in Seventeenth-Century England*, ix.

[22] Ernst Troeltsch, quoted in: G. Maier, *End of the Historical-Critical Method*, 93.

[23] Elmer Moeller, "The Meaning of Confessional Subscription," *Springfielder* 38:4 (1974), 204.

[24] David P. Scaer, "The Theology of Robert David Preus and His Person: Making a Difference," *CTQ* 74:1 (2010), 82.

can accept God's Word, and God Himself, without the Spirit. Faith is not a rational decision based on external evidence, but a divine gift which the Spirit works through external words. Certainty in His Word is a fruit of faith, not "circular reasoning," since only one in the Spirit can accept spiritual words. Unless the Spirit Himself convicts in the Word of God, man is left with only unreliable human testimonies. Both Luther and Calvin were thoroughly pre-modern in this regard—the claim of supernatural revelation did not trouble them rationally.[25]

What is worse is that there have been countless men arguing for the compatibility of these two methods for obtaining truth. But can an atheist hear a sermon or read God's Word as well as a believer?[26] When it comes to reading God's own Word, it is assumed that faith itself is a harmful presupposition. Scholarly tools and learning replace the Spirit for the modern.

This dividing line is fundamental, even when the conclusions seem within the bounds of historic Christianity. "The 'god' of higher criticism is no longer the God of Abraham, Isaac, and Jacob, but its own idolatrous creation."[27] It changes every aspect of Christian doctrine by distancing the Lord from history and truth from man. Nothing is out of bounds. Even if man restrains his critical judgment, he is the basis of that restraint, not the Lord's Word. In one Cartesian stroke, everything is desacralized and man-made. God is dethroned and man ascends in scholarly humility to pronounce judgment on all things, even the Maker Himself.

The Christian must "reject every interpretive methodology 'that uses some principle to set itself up as Lord over the text.'"[28] Divine authority which authenticates truth is at stake. If there is no eternal basis for theology, it becomes simply an exercise in spouting modernist heresies in pre-modern garb. Without the Spirit's grounding, there is no certainty. Theology becomes chameleon-like—mirroring modern philosophy, which is "primarily concerned with epistemology, the search for certainty."[29]

---

[25] "*Autopistia* [the self-convincing authority of Scripture] is not an exclusive Christian argument." D. P. Scaer, "Theology of Robert Preus," 85, 89.

[26] G. Maier, *Biblical Hermeneutics*, 44.

[27] E. F. Klug, "Review of *The End of the Historical-Critical Method*," 292–293.

[28] Hans Walter Wolff, quoted in: G. Maier, *Biblical Hermeneutics*, 320.

[29] Diogenes Allen, *Philosophy for Understanding Theology* (Atlanta: John Knox Press, 1985), 171.

Moderns are "always learning and never able to arrive at a knowledge of the truth" (2 Tim. 3:7).

Asking only what the Bible meant when it was written does not allow one to answer the question of what it means today. Methodology distorts the doctrinal content of Christianity. If you must choose between the unmutilated Bible and scientific facts, you will find your God.[30] "Our modern scientific-technological age involves what has been described as 'methodological atheism,' . . . so that science may be termed 'the Christian form of godlessness'."[31] Modernism is not against Christianity itself—it rejects every conception of the divine that might limit the critic. Its error is not in the results or final conclusions, but its atheistic foundations, so that modern "atheism is the vague feeling of uncertainty within their own minds."[32] As a result, the insistence on the binding force of the literal meaning of the Bible has become the ultimate superstition.

---

[30] William Edward Jelf, *Supremacy of Scripture: An Examination into the Principles and Statements Advanced in the Essay on the Education of the World* (London: Saunders, Otley, & Co., 1861; https://books.google.com/books?id=oWQXAAAAYAAJ), 141.

[31] Joseph C. McLelland, *Prometheus Rebound: The Irony of Atheism* (Waterloo, Ontario: Wilfrid Laurier University Press, 1988), 6.

[32] R. S. Westfall, *Science and Religion in Seventeenth-Century England*, 219.

Chapter 13

Modern A-Theology

A leading neo-orthodox theologian of the 20th century summarized, quite contentedly, the modern situation: "The breakdown of the doctrine of verbal inspiration as the result of modern scientific knowledge—both of natural science and of historical science—caused the collapse of the whole edifice of orthodox doctrine."[1] As the foundation of theology has been eroded, academic theologians continue to talk and write, doing the very thing God is supposedly unable to do. Yet under modern, Enlightenment assumptions, nothing concrete is asserted. "Semantical atheists" today believe that "no language about God can be cognitively meaningful."[2] Modern theology has become philosophy wrapped in theological language. Without an eternal Word of God, how can a man even speak of an eternal God? Outside of the scriptural teaching of inspiration, the historical ditch set up by modernism is impassable.

The loss of the absolute, unconditioned, eternally valid words of God leads to the denial of every aspect of Christianity. Modernism is an *a priori* assumption that unravels not only Scripture but all possible speech about God. Therefore, the Bible is basically human attempts to "reproduce, in human thoughts and expressions, this [unreproducible] Word of God in definite human situations."[3] The entire character of theology, without a sure Word of God, is a thoroughly human enterprise, with nothing to commend it. "The tragic history of twentieth century

---

[1] E. Brunner, *Revelation and Reason*, 11.
[2] Norman L. Geisler in *Inerrancy*, ed. N. L. Geisler, 333.
[3] Samuel Nafzger in *Studies in Lutheran Hermeneutics*, ed. J. Reumann, 111.

theology is the creation of God in our own philosophical or cultural image."[4] This has led to "the very term *God* as the ultimate non-sense word," so men conveniently "formulate their own creative statements . . . [to] play the ventriloquist" for this vocable.[5]

The question of the basis and nature of authority in theology cannot be answered scientifically. The problem with certainty is spiritual.[6] Faith has been untethered from truth and actual right and wrong, and therefore from absolute guilt and forgiveness. This has left it floating meaningless in a sea of theological vocabulary that says nothing concrete. It is devoid of facts. All that is left is human opinions, with no divine words as a lever to get underneath them. Timeless dogma is impossible in this line of thinking. Academic theology, then, is so often "a forum for what different men are saying *about* the text rather than a laboratory to listen to what the text, in fact, is saying."[7]

There is no absolute certainty apart from the truth given by God. If truth is framed in terms of scientific facts, theology becomes harmless and ultimately nonjudgmental. Under the critical mindset, "confessional differences fade almost completely into the background."[8] Reason, as the modern master and fount of theology, alters all doctrines. Luther's consistent denigration of reason and his "association of reason and fanaticism sounds quite extraordinary in *our* ears." According to Luther, "the source of all heresy is 'measuring doctrine by reason,' 'listening to the devil's bride'."[9] Fanaticism, that is, listening to reason, is the only acceptable method in modern times.

The scientific mindset is always critical. It doubts, which is the opposite of believing. So when the critic says anything about God, the speaker immediately becomes a mystic, an enthusiast, a fanatic.

> As long as we apply only the principles of historical criticism to the words of Holy Scripture, we might well find in the end that all we have is a heap of theological statements which, in the last analysis, contradict each other. Only when we

---

[4] J. W. Montgomery, *Crisis in Lutheran Theology*, 37.

[5] Carl F. H. Henry, *God, Revelation, and Authority: God Who Speaks and Shows: Fifteen Theses, Part Two*, vol. 3 of 6 (Waco, TX: Word Books, 1979), 404, 407.

[6] *Inerrancy*, ed. N. L. Geisler, 344.

[7] Concordia Seminary student letter to President J. A. O. Preus (Jan. 21, 1971). P. A. Zimmerman, *A Seminary in Crisis*, 403.

[8] K. Scholder, *Birth of Modern Critical Theology*, 6–7.

[9] B. A. Gerrish, *Grace and Reason*, 28.

really listen to what the Scriptures say shall we be free from the danger of wanting to force individual statements into some over-all system.[10]

Either God is the author of Scripture and doctrine, or man makes up both. For modernism, "in contrast to the Word of God, the subject and author in the case of doctrine is always man."[11]

God's Word is truth, but truth is more than exactness and scientific jargon. "Inerrancy does not demand the technical language of modern science. One should not expect the writers of Scripture to use the language of modern scientific empiricism."[12] Scripture also lacks modern verbal precision in citation. The lack of references and footnotes for sources troubles the modern, but the Spirit, as the Bible's author, is free to use His own words and actions truthfully, even when it seems contradictory to man. As the author of the Old Testament, He interprets His Old Testament words in the New Testament as He sees fit.

To admit error in an actual revelation is to put oneself above God—hardly a Christian virtue. Moderns are satisfied with a precise materialistic explanation that says very little. To be biblical and normed by its clear words is the worst theological sin today. In modern criticism,

> we are repeatedly presented with highly rationalized suppositions, built layer upon layer into intriguing structures of marvelous intricacy. But when we look for evidence, there is very rarely anything which would be convincing, at least to leading literary historians in the humanities. It is a pity to see eminent scholarly minds spending so much time on such elaborate intellectual jigsaw puzzles.[13]

It is teaching without divine authority.

"It is the settled conviction that the traditional conceptions of revelation are incompatible with modern developments in history and science."[14] Revelation is often styled as "self-revelation" by moderns, but "self-revelation," a throw-away word, is the counterpart to verbal inspiration. Revelation means communication. How does God's "self"

---

[10] James Arne Nestigen, quoted in: Eugene F. Klug, "Word and Scripture in Luther Studies since World War II," *Trinity Journal* 5 (1984), 10.

[11] G. Ebeling, *Word and Faith*, 175.

[12] Paul D. Feinberg in *Inerrancy,* ed. N. L. Geisler, 300.

[13] Albert C. Outler, quoted in: H. P. Hamann, "A Plea for Commonsense in Exegesis," *CTQ* 24:2 (April 1978), 128.

[14] W. J. Abraham, *Divine Revelation*, 1.

come to us without words? To grasp the bare nature and essence of God Himself is pure mysticism. In anti-revelation theology God has nothing at all to do with man rationally, but somehow man appropriates the entire substance of God in a personal encounter. This is neither Christian nor scriptural. The Bible forbids seeing God face-to-face, since He is a consuming fire to rebellious, unholy sinners (Heb. 12:29). Instead, "God reveals Himself and His will" in verbal communication.[15]

Critical examination endows man with divine authority to make all "witnesses" bow before him, even divine utterances. This is to be rejected, because man is not the highest authority. What is above the Word of God? One can only believe or reject the highest authority, not correct it or select the most appealing parts. "The theologian does not treat *God in Himself,* but *God in His revelation.*"[16] To deny the possibility of a revelation in words is atheism or some sort of inner mysticism.

This atheistic approach to theology renders quite small the theological problems of Luther's day, though there were some precedents for modernism in the enthusiasts who denied the sacraments. The Reformation argument of tradition and the papacy versus Scripture alone are all in the same fundamentalist category in the modern view. They all believed in inspiration and revelation, though there was strong disagreement about where it was located and how it was to be interpreted. However, all doctrine was considered divine, timeless, and above criticism. The modernist Rudolf Bultmann, at least, is consistent in his position: "to speak of God is simultaneously to speak of oneself."[17] The critical principle means that "theology is secular like every other academic subject."[18]

---

[15] T. H. Horne, *An Introduction to the Critical Study and Knowledge of the Holy Scriptures,* 2.

[16] Bernard Ramm, *Special Revelation and the Word of God* (Grand Rapids: Eerdmans, 1961), 14.

[17] H. Jackson Forstman, quoted in: J. W. Montgomery, *Crisis in Lutheran Theology,* 47.

[18] E. Brunner, *Revelation and Reason,* 391.

Chapter 14

Basis for the Word of God

"The central problem for twentieth century theology is its own epistemological basis. From what fountainhead does theology acquire her information . . . for preaching the Gospel?"[1] The scientific method relies on observable evidence and critical examination—which denotes a specific relationship to that which is studied. "Man, then, always stands over against when he observes; he is always himself a theme."[2] This "scientific neutrality" pits the truth in man against all external voices of authority. It is rightly said that inspiration is "unscientific. It is an abstract, *a priori* affirmation, not resting on objective [man-centered] facts. . . . "[3] Inspiration speaks of God's voice, which does not allow scientific objectivity at all. It demands submission and belief, which are intimately related in the Gospel. "The revelation of the mystery that was kept secret for long ages . . . has now been disclosed and through the prophetic writings has been made known to all nations, according to the command of the eternal God, to bring about the obedience of faith" (Rom. 16:25–26). Faith relies on doctrinal content about God's will, which He has disclosed in words.

Are the true facts of God established by God or by man? The issue is not so much Scripture itself, but whether man allows, in his critical judgment, the Lord to speak today. Scripture is not the only source

---

[1]C. H. Pinnock, *A Defense of Biblical Infallibility*, 1.
[2]H. Thielicke, *Prolegomena*, 35.
[3]H. P. Smith, *Inspiration and Inerrancy*, 32.

of truth, since "preaching predates Scripture."[4] The issue is the Word of God, which has taken many forms. In Old Testament times visions, dreams, angelic messengers, theophanies, and the direct "Word of the Lord" abounded. The Lord even wrote on a wall (Dan. 5). However, Scripture is the only revelation available to us today. Any attack on Scripture is an attack on God.

The basis for God's Word is the Spirit given through His Word. "The ground of testimony is the union of Word and Spirit." But Calvinists, unlike Lutherans, never held to a complete unity of Word and Spirit. They can speak of the "double structure" of revelation: the "inner and outer" Words.[5] Much like their conception of the Lord's Supper and in their teaching of the communication of attributes between the two natures of Christ, the spiritual is never identified completely with the physical. The modernist heresy is a more radical separation of Word and the Spirit.

In profane (secular) hermeneutics unbelief is supposedly unbiased. Everything, even God, is a hypothesis to be tested. Yet God either speaks or does not speak. There is nothing behind the Word which man may attain and pin-down for study. The human situations which "caused" the Scriptures and the historical situations which they describe do not hinder their application to any situation or time. The meaning of the human words in the pre-modern era "transcended the intentions of its human authors."[6] Inspiration grounds the authority of Scripture outside all men, even its human authors, in God Himself.

How can one know Scripture is God's Word and not man's? The certainty of divine authority is worked by the same Spirit who inspired the Word of truth. The "fullness of conviction" is "produced in the mind of the believer by the Holy Spirit."[7] The Spirit is present where the Word is heard. "The Holy Spirit is no Skeptic, and it is not doubts or mere opinions that he has written on our hearts, but assertions more sure and certain than life itself and all experience."[8] Belief in God's Word, worked by the Spirit, is the proper approach to that Word.

---

[4] B. Ramm, *The Witness of the Spirit*, 19.
[5] B. Ramm, *The Witness of the Spirit*, 17, 33.
[6] P. Harrison, *The Bible, Protestantism, and the Rise of Natural Science*, 125.
[7] B. Ramm, *The Witness of the Spirit*, 7.
[8] Luther, *The Bondage of the Will* (1525), LW 33:24.

"The entire theory that science is free from presuppositions rests upon the great, false presupposition that man can be without presuppositions."[9] Man is a sinner, under the wrath of the Holy God. If "reason alone judges the credibility and contents of Scripture," then he is more god-like than God Himself.[10] Does God justify man or does man verify God's words? The answer to this question affects all doctrines, especially the declaration of righteousness in the Gospel.

Reading Scripture is a spiritual endeavor that encompasses the whole life of the reader.

> Above all things it is quite certain that one can not search into the Holy Scriptures by means of study, nor by means of the intellect. Therefore begin with prayer, that the Lord grant unto you the true understanding of His Word. There is no interpreter of the Word of God, except the Author of the Word, God Himself.[11]

Man's spirit and understanding can only lead to death, because of sin. So, "Scripture is to be understood alone through that Spirit by whom it is written, which Spirit you can find more present and alive nowhere than in His Holy Scripture written by him."[12] Language skills and a critical mind are not the keys to interpreting Scripture. "The goal of exegesis cannot be reached without the actual presence of the Spirit."[13] The critical method does not support divine faith. It offers a man-made salvation.

"The right of private interpretation," which ignores the value of tradition, is, in essence, a scientific prejudgment.[14] It is centered on man as the critical apprehender of all truth, apart from past and present human authorities. All old authorities and creeds are disregarded as meaningless tradition, carrying no weight. The "Bible-believer" is free to see the Bible as the sole object of *his* study, looking at nothing else. This scientific activity appeals to the arrogant modern spirit, albeit to one who is willing to criticize the faith-based denial of the supernatural. So, the Bible-believer is more scientific in at least taking Scripture on its own terms, but the disregarding of old witnesses is decidedly modern. This

---

[9] F. Bettex, *The Bible the Word of God*, 222.

[10] H. D. McDonald, *Theories of Revelation*, 1:110.

[11] Luther, letter to Spalatin, quoted in: F. Bettex, *The Bible the Word of God*, 227.

[12] Luther, quoted in: J. M. Reu, *Luther and the Scriptures*, 76.

[13] D. M. Farkasfalvy, *Inspiration and Interpretation*, 129.

[14] J. Y. Simpson, *Landmarks in the Struggle between Science and Religion*, 139.

individualistic thinking has been more prevalent in Reformed thinkers than Lutheran thinkers, historically. Where the Spirit has given wisdom, it should not be ignored. The Catalog of Testimonies, an appendix to the Book of Concord, displays the thorough use of the church fathers by early Lutherans, though critically. This shows that the modern spirit is at odds with true Lutheranism.

Accepting the Word of God is not a rational decision. The critic is not above God, that he may reject the Word without rejecting the Author of the Word. The rationalist, who denies the Spirit's role in authorizing His own Word, reasons thus: "The decision to believe, to have faith in God's Word is comparable to the marriage decision, i.e., you accept and weigh all the evidence you can and choose."[15] God cannot be accepted any more than man can create Himself. This assumed neutrality is entirely anti-scriptural—God's own witness is made equal to all human sources in the decision-making process. To make God's Word initially equal to man's word is unbelief, a rejection of the Spirit in the Word. Only Christ can free the sinful mind in its rebellion and unbelief. "The recognition of the divine authority of Scripture is a fruit of faith."[16]

Moderns are troubled by the difficulties and logical gaps in Scripture. The impossibility of rationally harmonizing the Scriptures is well known. In the scientific acceptance of Scripture, it becomes necessary to completely answer all objections before continuing on to verify it. Parallel accounts do differ, but they cannot be proved to contradict one another—they are neither precise nor full enough for man to sit and judge the presumed unity of Scripture. The properties of the Word of God, including its clarity, sufficiency, and unity, are all articles of faith. Moderns assume clarity in man's reason, but the pre-modern (and biblical) approach acknowledges the reality of sin:

> The words of the Testament are in themselves very perspicuous, but are variously interpreted; because many, neglecting the literal and proper sense, studiously seek a foreign one ... because of the perverseness or imbecility of men. The obscurity which lies in the subject [man] must not be transferred

---

[15] Martin R. Noland, "Walther and the Revival of Confessional Lutheranism," *CTQ* 75:3–4 (2011), 214.

[16] Ralph A. Bohlmann, *Principles of Biblical Interpretation in the Lutheran Confessions* (St. Louis: Concordia Publishing House, 1968), 131.

to the object [Scripture].[17]

Man lives by faith in the Word, not by critical reason. Luther was concerned with the literal accuracy of parallel accounts, but warned not to go beyond the text itself:

> The evangelists do not all observe the same chronological order. The one may place an event at an earlier, the other at a later time. Mark, too, chooses the day after Palm Sunday for this story [of the cleansing of the temple]. It may also be that the Lord did this more than once, and that John reports the first, Matthew, the second event. Be that as it may, whether it happened sooner or later, whether it happened once or twice, this will not prejudice our faith.[18]

Luther let the plain words of the accounts stand together without contradiction, exactly as they have been revealed.

To hear God's Word, even in derivative forms, such as teaching and preaching, is to hear God Himself. Luther was no rationalist. He describes David's blessed trust in the Messiah:

> I am sure and convinced of this not only because this has been promised me by God, whose words are certain and reliable and will not lie to me, but also because . . . I implicitly trust God's Word with all confidence. Therefore I am cheerful and stand ready to live or to die when and how God wills.[19]

This is exactly how Scripture reads: God has spoken, so the matter is more firm than all creation. "It is written" in the New Testament carries the highest authority, the very same as "Thus saith the Lord," which is found some 400 times in the Old Testament. The Bible recognizes the human authors and also affirms that their words are fully the undiluted words of the Spirit. "For the most part the [New Testament] quotation formulae which refer to the [Old Testament] human authors are freely interchangeable with those which refer to the divine subject."[20] "In Hebrews the Spirit is continually named as the speaker in the Old

---

[17] Heinrich Schmid, *The Doctrinal Theology of the Evangelical Lutheran Church*, trans. Charles A. Hay and Henry E. Jacobs (Minneapolis: Augsburg, 1875; reprint, Philadelphia: United Lutheran Publishing, 1961), 73.

[18] *Sermons on the Gospel of John* (1530–31), LW 22:218.

[19] *On the Last Words of David* (1543), LW 15:271–272.

[20] *Theological Dictionary of the New Testament,* ed. Gerhard Kittel, ed. and trans. Geoffrey W. Bromiley (Grand Rapids: Eerdmans, 1964), 4:111.

Testament passages."[21] This particular book cites no human author, except "someone" in 2:6. The Bible's authority is inherent in the inspired words, even if widely misused and disregarded by sinners.

The basis of God's Word is not man's faith or acceptance, but the Spirit's testimony worked via the Word. The ultimate authority for a word is its speaker. The centurion, with faith not found in Israel, said to Jesus: "only say the word, and my servant will be healed. For I too am a man under authority, with soldiers under me. And I say to one, 'Go,' and he goes, and to another, 'Come,' and he comes, and to my servant, 'Do this,' and he does it" (Mt. 8:8–10). The Word of Jesus possesses unlimited authority.

The denial of the role of the Spirit appears quite close to a denial of the need for the Spirit. The Word of God authenticates itself. Man cannot validate God's words. All reservations in accepting God's Word are due to a lack of faith in the God who speaks. This applies to all critical methodology. "To the [old theology] all historical questions were really dogmatic."[22] In orthodoxy, the truth of Scripture guarantees history above all doubts of man.

"If God does not speak, it is not clear how we can know in any substantial way His intentions and purposes."[23] Man jumps at the chance to vindicate God and His Word, but to verify or authenticate it can only make it less than a divine Word. "God alone is the appropriate and genuine witness" to Himself.[24]

---

[21] Lewis Sperry Chafer, *Systematic Theology*, 8 vols. (Dallas: Dallas Seminary Press, 1947), 1:71.

[22] A. M. Fairbairn, *The Place of Christ in Modern Theology* (London: Hodder and Stoughton, 1893), 3.

[23] W. J. Abraham, *Divine Revelation*, 25.

[24] Johann Gerhard, quoted in: Adolf Hoenecke, *Evangelical Lutheran Dogmatics*, 3 vols., trans. James Langebartels and Heinrich Vogel (Milwaukee: Northwestern Publishing House, 2009), 1:451.

Part V

Modern Anthropology: Reverse Arianism

Chapter 15

Assumed Incompatibility of Man and God

In contemporary thinking, "objective authority is a transgression of man's fundamental freedom and corrupting of his rightful autonomy."[1] Therefore the doctrine of inspiration has become a plague and the greatest Christian mistake to moderns. But what does the "errancy" of Scripture confess? "The doctrine of errancy clearly implies that God cannot so interact with what is human in a way that He preserves it from all error without destroying its essential humanness."[2] At the root of the inerrancy debate are fundamentally different conceptions of God and man, and their compatibility. Modernism is not simply a differing interpretation of doctrine, but a super-Arminianism that denies all divine involvement with man, by giving all supernatural powers to man.

The Spirit's inspiration is pejoratively called "mechanical" to discredit it. Behind the word "mechanical" is a vast unscriptural conception of God. "Recent theology found that the humanness of the writers is inconsistent with inerrancy."[3] "If a man were given the power to utter the eternal wisdom of God and write down absolute truth, he would be unmanned, dehumanized."[4] Man is supposedly violated by God's

---

[1] H. D. McDonald, *Theories of Revelation*, 2:332.
[2] Harold O. J. Brown in *Challenges to Inerrancy,* eds. G. R. Lewis and B. A. Demarest, 390.
[3] Gordon R. Lewis in *Inerrancy,* ed. N. L. Geisler, 229.
[4] T. Engelder, "Verbal Inspiration: A Stumbling Block," Part 6, 202.

interaction—an assumption positing an impassable divide between them. This is nothing less than an "absolute separation of time and eternity," which shreds all of Christianity.[5]

Moderns generally "assert that finite human language is incapable of expressing the infinite."[6] It is not just language that is weak, but God's supposed distance and absenteeism make the idea of Him contacting man incredible. Those with a closed, materialistic view of the universe say: "inspiration . . . does not change the finite into the infinite."[7] This is a false application of natural philosophy, which makes God bow down to reason and its fleshly conclusions.

God's involvement in the creation He upholds supposedly makes man inhuman. This is a very strange definition of humanity—one that actually conflicts with the Bible's definition of God. Reason is methodologically assumed to be "the candle of the Lord," but it is not now so directly stated as it was at the beginning of the Enlightenment.[8] Man's authority to be a god-like critic can only be preserved by declaring man off limits to God. Modernism is not concerned with God, but its assumptions about man make traditional doctrines of God untenable.

Refusing to identify Scripture and the Word of God is "rooted in the bedrock conviction of the mutual exclusiveness of the divine and the human."[9] Owing to scientific assumptions, this is much more deist than Christian. "The objection against revelation that it makes man a mere machine," is traced to the early 1700s—to the beginnings of the Enlightenment period.[10]

In the pre-modern conception, "divine inspiration does not destroy individuality, it elevates it."[11] In modern thought God is an infinite idea with no real connections to this world. "The apostles and prophets . . . are finite human beings, they are simply incapable of utterly divine words."[12] This is pagan philosophy disguised as Christian theology.

---

[5] J. W. Montgomery, *Crisis in Lutheran Theology*, 24.
[6] G. Goldsworthy, *Gospel-Centered Hermeneutics*, 34.
[7] H. P. Smith, *Inspiration and Inerrancy*, 38.
[8] H. D. McDonald, *Theories of Revelation*, 1:42.
[9] Henry Krabbendam in *Inerrancy*, ed. N. L. Geisler, 440.
[10] T. H. Horne, *An Introduction to the Critical Study and Knowledge of the Holy Scriptures*, 25.
[11] F. Bettex, *The Bible the Word of God*, 303.
[12] P. H. Nafzger, *These Are Written*, 63.

Assumed Incompatibility of Man and God

There is no way to prove it—Scripture certainly cannot be cited. This understanding is rooted in the fundamental conviction that God and man are so equal in powers that they must be placed in opposition. It follows in the steps of the Calvinist denial of the Lord's Supper, but goes much further than that and concludes that God cannot use human language, nor can His will be circumscribed by words. Only an inconsistency in this philosophy would countenance the incarnation of Christ—the closest joining of God and man possible.

The following statement from around 1800 shows that inspiration as pure revelation is the point most attacked by modern intellectual atheism: it is "impossible for God himself so to inspire a man as to preserve him from error without destroying his [human] nature."[13] Because the mere possibility of God's Word is attacked, this thinking cannot be dismissed with a simple citation, like: "With man this is impossible, but with God all things are possible" (Mt. 19:26). In fact, divine knowledge is no longer allowed by critical man. The modern definition of man denies competing authorities. To be human is no longer to err, but to think divine thoughts and determine what is possible for God.

The neo-orthodox theologian Emil Brunner called the "traditional equation of the 'word' of the Bible with the 'Word of God' . . . a breach of the Second Commandment: it is the deification of a creature, bibliolatry."[14] Instead, he talks of "self-revelation" and "self-disclosure," as if man can appropriate the bare substance of God apart from Christ's Word. All "self" talk is dealing with God's inner nature apart from Christ, which is not for us to know.

To preserve man's new-found critical authority, all rational, propositional, and communicative links to God are broken. "God must be denied to guarantee man's freedom" in modernism.[15] A modern can spout all the God-talk he wants—it is his assumptions about man that destroy his theology.

The modernist theologian begins with a false antithesis: "if these men [the biblical writers] could say and write only what God wanted,

---

[13] Andrew Fuller (d. 1815), quoted in: *Inerrancy and the Church*, ed. J. D. Hannah, 332.

[14] E. Brunner, *Revelation and Reason*, 120.

[15] C. Fabro, *God in Exile*, 26.

and were not free to make mistakes, then their freedom as humans was compromised or destroyed."[16] This false view of man was invented by the god of reason. That the Word of the Lord comes to man and that the Lord speaks are integral to the defining the scriptural God. There is the most beautiful intimacy in God speaking and, in response, man praying in the Spirit to "our Father." The belief in verbal inspiration, describing absolute divine revelation, certifies the Bible's preeminent authority. Yet, "this type of Protestantism is now almost extinct amongst the educated."[17]

The Spirit carried the hearts and minds of the biblical writers already converted and moved by the Spirit. "God uses, rather than annuls their wills."[18] The Spirit chose believers and those who trusted in Christ, which is instructive. The command to write does not rule out external circumstances, but "establishes it," since the same Christ who arranges all things is one with the inspiring Spirit.[19]

"All infallibilities presuppose an idea of grace mechanically irresistible."[20] That is not how Scripture itself speaks. Through grace man is restored in Christ, so that he is not in slavery to sin. By faith in the Son he is brought closer to his intended state, one of fellowship with the Father. Submission to Christ is willing, not forced or mechanical. How can the love of the Father for His baptized children be robotic or impersonal?

An inerrancy which denotes authority is prohibited by moderns, but a dynamic and permissive inerrancy of purpose is praised. This goal-oriented truth does not rule out any assumptions about man's powers. "Functional inerrancy leads to the distinction between the form and content of Scripture, to the distinction between the divine and human aspects of Scripture."[21] The unity of the dual authorship is not just denied, but declared impossible in principle. This allows modern man

---

[16] S. Becker, *The Scriptures—Inspired of God*, 39.
[17] Edwyn Robert Bevan, *Sibyls and Seers: A Survey of Some Ancient Theories of Revelation and Inspiration* (London: G. Allen & Unwin, 1928), 40.
[18] L. S. Chafer, *Systematic Theology*, 1:68.
[19] Johann Gerhard, quoted in: A. Hoenecke, *Evangelical Lutheran Dogmatics*, 1:414.
[20] H. D. McDonald, *Theories of Revelation*, 2:289.
[21] William John Hausmann, *Science and the Bible in Lutheran Theology: From Luther to the Missouri Synod* (Washington, D.C.: University Press of America, 1978), 97.

to judge the human parts, although he has trouble exactly identifying the divine in Scripture.

Inerrancy, in modern thinking, robs the human authors of their freedom and humanity. The uplifting of man's rational powers and his independence were paramount concerns of the Enlightenment. Man who wants to be divine is limited by the necessity of God the Spirit, who lifts him out of his erring, lost state. Inspiration as revelation implies reason is insufficient and errant. This is a fight for total authority.

The modernist position puts God far away from men: "The Word of God itself is by definition beyond the reach of man."[22] "If we equate the Word of God with the Scripture, we are confusing heavenly things with things historical."[23] The roots of this complete separation are neither Lutheran, nor Christian. In the Reformation period the enthusiasts were the first to divide the human, outer word from the divine, inner Word. Luther harshly condemned all such fanaticism as

> the new error which placed beside and above the Scriptures an inner Word, revealing itself, as was claimed, within the heart of the individual believer. That man's own reason could, under any circumstances, ignoring the objective Word, . . . lead men to God or to the knowledge of divine truth was an idea excluded by Luther's doctrine [of sin].[24]

In Luther's verdict the search for a word beyond the written one is fanaticism. Its current institutionalization in the academy can be nothing but an all-encompassing heresy. He spoke to this issue directly by applying a classic passage on biblical inspiration to affirm a lower, derivative inspiration for every faithful pastor:

> I am sure and certain, when I go up to the pulpit to preach or read, that it is not my Word I speak, but that my tongue is the pen of a ready writer, as the Psalmist has it. God speaks in the prophets and men of God, as St. Peter in his epistle says: "The holy men of God spake as they were moved by the Holy Ghost" [2 Pet. 1:21]. Therefore we must not separate or part God and man, according to our natural reason and understanding. In like manner, every hearer must

---

[22] Henry Krabbendam in *Inerrancy,* ed. N. L. Geisler, 440.

[23] Joseph Sittler, quoted in: *Studies in Lutheran Hermeneutics,* ed. J. Reumann, 324.

[24] Julius Köstlin, *The Theology of Luther in its Historical Development and Inner Harmony,* 2 vols., trans. Charles E. Hay (Philadelphia: Lutheran Publication Society, 1863), 2:220.

say: "I hear not St. Paul, St. Peter, or a man speak, but God himself."[25]

The attack on the Word of God has snowballed into "widespread doubt as to whether language can convey transcendent realities at all."[26] The confession of the Spirit's speaking in the Nicene Creed became philosophically problematic: how can the Spirit speak through men? The modern question is: how can two authors produce one work without violating each other or resulting in disjointed communication? Moderns, following the method of doubt, are suspicious of this claim. Even if Scripture is verified after the scientific courtship, the authority resides with man. According to the Nicene confession of faith, however, "the writings are authoritative, not because of the human author, but because God is regarded as the ultimate author."[27] While the scientific method seems to be objective, that it is without philosophical assumptions, it has definite biases and tendencies.

Finiteness and infinity are logical, mathematical concepts. "All science as it moves toward perfection becomes mathematical in its ideas."[28] Pure and true knowledge, in modern thinking, has mathematical precision. The modern, while eschewing ultimate truth claims, is intensely interested in exactness and technical measurement. The numbers of scientific data ring true, though moderns usually confuse precision with accuracy. At face value, the rational, "natural laws" of mathematics seem obvious and unbiased to the modernist. But in assuming a coherent, rational system, scientific accuracy becomes a false religion when applied to divine things. After all, God is Father, Son, and Spirit in Scripture, not an impersonal infinity. "Modern thought from its very outset returned to the mathematical approach to reality, seeing in physico-mathematical sciences the most perfect and valid paradigm of knowledge."[29]

In the doctrine of inspiration, "1 author + 1 author = 2 authors" is a false equation. This truth is above and contrary to the fleshly mind. Mathematics is an abstraction, so that it is not directly connected

---

[25] Luther's *Table Talk*. Ray Comfort, *Luther Gold: Pure Refined*, ed. Mary Ruth Murray (Bridge Logos Foundation: Alachua FL, 2009), 95.

[26] James I. Packer in *Inerrancy*, ed. N. L. Geisler, 203.

[27] John W. Wenham in *Inerrancy*, ed. N. L. Geisler, 17.

[28] J. Y. Simpson, *Landmarks in the Struggle between Science and Religion*, 147.

[29] C. Fabro, *God in Exile*, 1122.

to reality. Under the approach of rational, natural religion the first doctrine to go is always the Trinity—as with the Socinians. Three divine persons which equate to one divine substance is contrary to the "laws" of mathematics. Mathematics derived from reason is incapable of describing the divinity—only God's words are a suitable basis.

The philosophical basis for science is widely ignored, but it certainly has one. "Modern science [is] that skillful commingling of mathematics with observation, hypothesis and controlled experiment."[30] It isolates items and their properties for observation, but does not handle relationships between diverse things. This is why science has trouble describing the one flesh union of marriage, or even friendships. Mathematical descriptions cannot see unity in what is separated by critical methods. Mathematics is not a divine language—it is a tool of reason, which does not grasp the divine. Luther's advice is still applicable:

> [David] wants to lay hold of the real teacher of the Scriptures himself [the Spirit], so that he may not seize upon them pell-mell with his reason and become his own teacher. For such practices gives rise to factious spirits who allow themselves to nurture the delusion that the Scriptures are subject to them and can easily be grasped with their reason, as if they were *Markolf* [a legend] or Aesop's Fables, for which no Holy Spirit and no prayers are needed.[31]

The naturalistic attack on inspiration is not only an attack on Protestantism, but Romanism, Eastern Orthodoxy, and all orthodox Christianity. The modern development of historical consciousness allows one "to get beyond the Scripture/Tradition debates."[32] But the liberation from authority is also a liberation from God and His truth. In trying to move past confessional divides and polemical wars, truth itself was put in the rear-view mirror. The fundamentalist upsets the atheistic peace of the modern world—where unity is rooted in each person's own intrinsic authority. Modern peace is not to be disturbed by an authority above man. Pluralistic modernists fight vehemently against any fundamentalism which asserts divine authority in some form. The new religion fights against the old versions.

"In the Hebrew religion you get an intimacy, a habit of conversa-

---

[30] A. Richardson, *The Bible in the Age of Science*, 16.
[31] *Preface to German Writings* (1539), LW 34:286.
[32] K. Hagen, *Foundations of Theology in the Continental Reformation*, 231.

tion between God and man that went beyond anything in the pagan Greeks."[33] It is Greek philosophy that views man as inactive and passive in its mantic conceptions of inspiration. "Plato taught that a man cannot become a prophet, nor can he utter an oracle until he has abandoned his own reason, and has allowed the Divine to occupy its place."[34] In confessing "two authors, undiminished in their respective roles and natures . . . there [is] no way of avoiding a [modern] tension."[35] Premoderns did not have the same difficulty in believing that God could speak plainly to men, through men.

The orthodox Lutheran Fathers are ridiculed by moderns for saying that God commanded the human authors of Scripture to write. But an impartial observer, not prejudiced against God's truth, will acknowledge that the command to write is in the Old Testament (Ex. 17:14, 34:27; Is. 8:1, 30:8). "The word that came to Jeremiah from the Lord: 'Thus says the Lord, the God of Israel: Write in a book all the words that I have spoken to you'" (Jer. 30:1–2). The God who made man and his ability to communicate is deemed incompetent to speak. This is a denial of the Creator who made communicating man. "Who has made man's mouth? Who makes him mute, or deaf, or seeing, or blind? Is it not I, the Lord? Now therefore go [Moses], and I will be with your mouth and teach you what you shall speak" (Ex. 4:11–12).

Because man and God are not allowed to mix, their contact point, the Word of God, is attacked. This is a major source of modern uneasiness with identifying God as the full author of Scripture. Consider a description of Aquinas' philosophy: in "biblical inspiration the Absolute Divine Cause fully and totally dominates the instrument by causing it to cause what God intends, while at the same time allowing it to remain a free human agent."[36] This freedom is an impossible slavery to the critical modern. In the modern separation of God and man, even faith itself is made impossible, for "Christian faith can exist only *vis-à-vis* the Word of God."[37] Faith does not depend on Scripture, but it is created by a word of God. God's Word is more than man's word

---

[33] E. R. Bevan, *Sibyls and Seers*, 103.

[34] A. A. Zaun, "A Study of the Idea of the Verbal Inspiration," 7.

[35] D. M. Farkasfalvy, *Inspiration and Interpretation*, 182.

[36] D. M. Farkasfalvy, *Inspiration and Interpretation*, 150.

[37] Francis Pieper, quoted in: T. Engelder, "Verbal Inspiration: A Stumbling Block," Part 16, 901.

spoken loudly—it has a divine source outside of creation and man's dominion.[38] Modernism is much broader than an attack on Scripture. It is a muzzling of God Himself, so that man is unrestricted.

"It is odd to speak of a person simply revealing, in the same way as we speak of a person writing or walking or whistling."[39] The verb 'reveal' should have an object—the thing that is revealed, or given, to another. But the word "revelation" has been made meaningless: "Revelation is a personal act which takes place face-to-face."[40] One does not present one's substance, and neither does God, who is a consuming fire. The Word is the covering of God, not as a replacement for the incarnate Christ, but it is the way in which Christ comes to individuals. This is not rational to man in the least, that weak words of Christ should be the only way to God, yet true Lutheranism dared to follow the text to its fullest conclusions and connect God and man in the closest fashion in Christ. In contrast, it is the Reformed who still call the Lutheran doctrine of the ubiquity of Christ's human nature "patently incompossible," meaning "not capable of joint existence."[41] For those who confess that the human nature of Christ possesses the fullness of the majesty of the Godhead, human language is no obstacle.

Dual authorship does not make sense to modern Lutherans: "it is impossible, according to psychological criteria, for the same book to have two authors in any real sense."[42] The teaching that the Spirit's Scripture had two authors, one God and one man, is not conflicting in the least. "Scripture was not brought forth by the will of men, nor even by the co-operation of men."[43] This is God the Spirit's own testimony: "For no prophecy was ever produced by the will of man, but men spoke from God as they were carried along by the Holy Spirit" (2 Peter 1:21). This is the same bedrock as salvation by faith: complete monergistic action by the Spirit, who converts and illumines man. A synergistic idea

---

[38] A Christian should disagree with this statement: "faith in Christ necessitates the inspiration of Scripture," since many Old Testament saints before Moses believed in the Messiah without any Scripture.

[39] W. J. Abraham, *Divine Revelation*, 11.

[40] John Baillie, *The Idea of Revelation in Recent Thought* (New York: Columbia University Press, 1956), 127.

[41] R. A. Muller, *Post-Reformation Reformed Dogmatics*, 244.

[42] Traugott H. Rehwaldt, "The Other Understanding of the Inspiration Texts," *Concordia Theological Monthly* 43:6 (1972), 361.

[43] R. D. Preus, *Inspiration of Scripture*, 54–55.

of inspiration puts man on the same level as God, creating a tension where none exists, hence all "mechanical" and "docetic" language. In the modern view man must have contributed something non-divine in the writing process. Yet if God did not supply all the words and thoughts, with man willingly yet passively cooperating, the result would not be God's actual words. It is noteworthy that God chose to use holy, spiritual men, those who already believed by the Spirit, to write His words. Likewise, believers are not forced mechanically to do good works or believe.

"Inerrancy almost invariably results in a docetic view of the Bible."[44] The implication is that "God's words" would be a contradiction, since they cannot be truly human—like the teaching of docetism which said that Christ only appeared to be in the flesh. God's direct interaction becomes a transgression of man's law. The result is that "the humanness of the Bible . . . was virtually deified."[45] A fully divine Bible cannot be studied scientifically. Critical methods were used to "prove" God's distance from man's domain. Freedom became freedom from God and His "mechanically controlling" Spirit. How utterly un-Christian to assert that freedom and independence from the Spirit of God is truly human. "Now the Lord is the Spirit, and where the Spirit of the Lord is, there is freedom" (2 Cor. 3:17).

This "humanness" of the Scriptures is established by naturalistic observation. Visible experience rules instead of faith. "Docetic" is a modern code word for absolute authority that bears down on the critic and prevents scientific judgment. The word "mechanical" is actually a euphemism for the unthinkable pre-modern submission to direct revelation. A verbally inspired Scripture is "mechanical" because it cannot be critiqued or questioned. This divine legal authority demands that man simply accept it. No questioning or verification is possible. Hearing God's words cannot be a scientific activity, since the Spirit works through His words to bring saving life.

The paradox of inspiration is not in a different category than that of the doctrine of conversion. Good works are motivated by the Spirit, yet done willingly. Without the Spirit, though, man can only disbelieve

---

[44]Martin Scharlemann, quoted in: J. W. Montgomery, *Crisis in Lutheran Theology*, 35.

[45]H. D. McDonald, *Theories of Revelation*, 2:216.

and sin. How can one do good works only under the Spirit's influence, but do so willingly? This is the scriptural doctrine: the Spirit is given freely to free us, that "we might understand the things freely given us by God" (1 Cor. 2:12).

The modern objection to verbal inspiration is made to preserve man's freedom. God's movement of the authors is termed a "mechanical violation." Scripture actually speaks of this rational blindness: "no one comprehends the thoughts of God except the Spirit of God" (1 Cor. 2:11).

In divine monergism, God is the author and source of all good in man. Without this teaching, faith in any biblical sense is impossible. To deny the possibility of two authors for the one Word of God is to separate God from man. Inspiration is connected to the Spirit's role in the Christian life—"How is it then that David, in the Spirit, calls him Lord" (Mt. 22:43)? David spoke in, or by, the Spirit—the same Spirit of Christ that caused him to believe and confess Christ's righteousness in words: "that you may be justified in your words and blameless in your judgment" (Ps. 51:4).

In fact, without this Spirit no one can believe or confess Christ: "no one can say 'Jesus is Lord' except in the Holy Spirit" (1 Cor. 12:3). A denial of inspiration, based on a deified anthropology, seems inevitably to lead to a denial of justification. Man is forgiven by divine faith—which is completely God's doing—not a partly human faith. If man can believe without the Spirit, the scriptural Christ is not the object or author of that faith.

> It is correctly said that in conversion God, through the drawing of the Holy Ghost, makes out of stubborn and unwilling men willing ones, and that after such conversion in the daily exercise of repentance the regenerate will of man is not idle, but also cooperates in all the works of the Holy Ghost, which He performs through us.[46]

Truth, and even Christ's incarnation, hardly seem possible if Christ cannot bridge the chasm of human language. How is a true union of the divine and human natures of Christ even allowable in this scheme, if it cannot be rationally communicated? The scriptural Christ—the true God—taught and preached. Were His words also limited? The

---

[46] *Formula of Concord* (FC) Ep II:17; *Concordia Triglotta* (1921). Unless otherwise noted, all quotes from the Lutheran Confessions are from this edition.

philosophical separation of God and man in the denial of inspiration, if consistently carried through, would disqualify the personal union of Christ and every divine interaction with man. At the very least, it would be impossible to describe, preach, or confess Christ in human words, making the heart of Christianity inaccessible and practically useless.

Classic Christian labels and terminology lose almost all meaning in the face of this paganism. It attacks all churches and traditional approaches to theology. Modern historical and philosophical assumptions distance not just Scripture, but Christ Himself. In this modern view God can have nothing definite to do with man. The Maker of time has been divorced from history and declared separate for all eternity. The assertion of the priority of reason, and, accordingly, the new doctrine of man, is truly intellectual idolatry and "noetic apostasy."[47] In this light, "today [fundamentalism] means, simply, religious," that one is not a modern noetic, or mental, idolater.[48]

---

[47] G. Goldsworthy, *Gospel-Centered Hermeneutics*, 76.
[48] Leon Wieseltier in *The Fundamentalist Phenomenon,* ed. N. Cohen, 192.

Chapter 16

Super-Pelagianism

While "reason was generally [previously] regarded as incompetent in the religious sphere," the "genius of the modern era" is that no "sacrifice of the intellect" is needed to speak of God.[1] Scripture as an authority has been pushed aside for unrestricted scientific criticism. But to critique divine matters there must be a fulcrum, something to get the critic underneath the divine Word. The lever is the ability of man, so that modernism is synonymous with the denial of original sin. This new Enlightenment doctrine of man's goodness causes a fundamental reorientation of theology. "Near the heart of every theological debate since the Enlightenment there has been a question, either implicit or explicit, about the relation between [man's] freedom and [God's] authority."[2]

Socinus in the late 1500s taught that "a person can without the illumination of the Holy Spirit grasp everything [in Scripture] correctly."[3] His teaching "raised the underlying question of the competence of reason in theological matters."[4] In the pre-modern era, "the fact of revelation admitted excludes, *ipso facto*, the sufficiency of reason."[5] An unchecked emphasis on reason's powers makes the possibility of revelation dubious

---

[1] Frederick C. Beiser, *The Sovereignty of Reason: The Defense of Rationality in the Early English Enlightenment* (Princeton: Princeton University Press, 1996), 5; W. J. Abraham, *Divine Revelation*, 7.
[2] Allan D. Galloway in *The Science of Theology,* ed. P. Avis, 350.
[3] *Racovian Catechism,* quoted in: A. Hoenecke, *Evangelical Lutheran Dogmatics*, 1:485.
[4] R. A. Muller, *Post-Reformation Reformed Dogmatics*, 239.
[5] W. E. Jelf, *Supremacy of Scripture*, 70.

in modern thought. As a consequence, the word "inspiration" has been twisted to mean anything but God's mode of direct revelation. The existence of a special revelation restrains man's inherent potential and achievable knowledge. It stipulates that without God's illuminating Spirit and words man is incompetent.

To be without presuppositions is to come before God without Christ's righteousness—as an equal. "The origin of historical criticism is closely linked with the doctrine of human goodness," or approaching God in utter blasphemy.[6] Because of Adam's sin, "our evil heart is prejudiced against God's Word."[7]

The ancient heresy of Pelagianism is the false doctrine that man can naturally speak of and know divine things. "The rise of Pelagianism was part and parcel of the new rationalism," since man's rational powers were emphasized.[8] The first Enlightenment rationalists fought for the divine right of criticism, but this Pelagianism is an unstated assumption today. Modern theology talks positively of Christ with great verve, but ignores the biblical teaching that we are "dead in our trespasses," including intellectually (Eph. 2:5).

Modern attempts at reconciling the Scriptures with scientific authority make it hard to see them as divinely authoritative. One heretic of the mid-1600s claimed that by his efforts the "position of faith is again reconciled with natural reason."[9] The orthodox response was stark: "I declare that no more pernicious heresy has ever been circulated in the church than this doctrine."[10] In this the Roman Catholic, Reformed, and Lutheran polemicists were in agreement. Their united response is known today as fundamentalism, only because it is now the minority opinion.

Faith given by the Spirit is unscholarly. Knowledge of the biblical languages and the reading of many human authors are not substitutes for the Holy Spirit. To approach God, which can be done only through His Word, without the assumption of His enlightenment is un-Christian.

---

[6] G. Maier, *Biblical Hermeneutics*, 295.

[7] T. Engelder, "Verbal Inspiration: A Stumbling Block," Part 16, 917.

[8] F. C. Beiser, *The Sovereignty of Reason*, 18.

[9] Isaac La Peyrère, *Prae-Adamitae* (1655), quoted in: K. Scholder, *Birth of Modern Critical Theology*, 88.

[10] Johannes Micraelius (d. 1658), quoted in: K. Scholder, *Birth of Modern Critical Theology*, 87.

It is a form of Pelagianism that would make Pelagius blush. To talk about God without claiming the necessity of the Spirit is enthusiasm. It is to consider oneself on par with the Lord who dwells in glory. Even worse, faith and the Spirit Himself are counted as presuppositions, by definition harmful to scientific knowledge.

God and man have been completely split apart by the rights arrogated by man. Man, declared to be the highest authority, lowers the authority of Scripture. "Reason is an infallible guide in judging biblical matters"[11] "Reason, aided by experience and strengthened by wisdom and intellectual discipline, is our standard."[12] The effects of this assumption are incalculable, even if not stated in Enlightenment terms today. "In fact that amounts to making the human understanding God . . . to deny the fall of first man and the subsequent blindness of human knowledge . . . to deny the state of sin, the necessity of rebirth."[13] The unstated doctrine of man in modern theology undoes all the stated doctrine of God, no matter how eloquently spoken.

With no limitations on reason's subject matter or its ability to perceive ultimate truth, this new doctrine of man is described as "super-Pelagianism," "an almost boundless confidence in the capacities and possibilities of reason."[14] If reason without the Spirit can apprehend God's communication, there is also a "tension-free depiction of reason and revelation."[15] The assumed "human side" of Scripture is the crack into which the reader inserts himself and exercises authority over the whole. "The general presupposition that the Scriptures are the Word of God is in effect negated by the insistence that the 'human side' of the Scriptures necessarily makes them subject to every human limitation."[16] A pure, divine revelation becomes unthinkable and preposterous for man trusting in his newfound rational salvation. "The term inspiration

---

[11] This was the position of Jean Le Clerk expressed in 1685. Ronald R. Feuerhahn in *The Bible in the History of the Lutheran Church,* ed. J. Maxfield, 50.

[12] Charles Watts, "The Secular and Christian Standards of Morality," *Freethinker,* Sunday Nov. 3, 1895 (https://books.google.com/books?id=ilsvAAAAYAAJ=PA692&dq), 692.

[13] Nicolaus Arnoldus (d. 1680), quoted in: K. Scholder, *Birth of Modern Critical Theology,* 117.

[14] K. Scholder, *Birth of Modern Critical Theology,* 117, 115.

[15] G. Maier, *Biblical Hermeneutics,* 293.

[16] "Report of the Synodical President to the LCMS" (1972). P. A. Zimmerman, *A Seminary in Crisis,* 292.

loses its meaning when an attempt is made to divide it between God and man."[17] This is the case today, where inspiration is not so much denied as ruled nonsensical and outmoded by default. The modern Lutheran Herman Sasse tried to avoid "drowning the human side of Scripture in . . . over-emphasis on the divine."[18] Pre-moderns saw no such tension between God and man in the writing of Scripture.

"Pelagianism is the most reasonable (in the sense of rational) of heresies."[19] It requires no doctrinal presuppositions describing sin and the rebellion of man. The Lutheran Confessions "condemn the Pelagians and others, who teach that without the Holy Ghost, by the power of nature alone, we are able to love God above all things."[20] Yet in making doubt the absolute principle of knowledge, the first moderns were countered by the orthodox theologians: in "the orthodox replies, we continually find as one of the main charges that the Cartesians have exchanged the roles of God and man, of creator and creature."[21] Modernism is in essence the "gradual transference of divine attributes to human beings."[22]

The new doctrine of man was not billed as theological, but this was the unstated assumption of the philosophy of natural science. Scientific understandings of the world expressed strong religious assumptions in their "objective" natural laws. The idea of natural laws "bestows great powers upon human reason."[23] When ultimate truth is discoverable without the help of the Spirit and Scripture, man's status in relation to God changes. "In theological terms the new spirit is often evaluated as a kind of super-Pelagianism."[24] Reason is granted divine authority to

---

[17] Robert Haldane, *Authenticity and Inspiration of the Holy Scriptures Considered* (Edinburgh: John Lindsay, 1827), 69.

[18] Jeffery J. Kloha in *Scripture and the Church: Selected Essays of Herman Sasse*, eds. Jeffrey J. Kloha and Ronald R. Feuerhahn (St. Louis: Concordia Seminary, 1995), 349.

[19] Alister E. McGrath, *Intellectuals Don't Need God and Other Modern Myths: Building Bridges to Faith through Apologetics* (Grand Rapids: Zondervan, 1993), 150.

[20] *The Augsburg Confession* (AC) XVIII:8.

[21] K. Scholder, *Birth of Modern Critical Theology*, 117.

[22] This is "why we do not understand why radical Islam sees our liberal world as impious and immoral." It is by pre-modern standards. Michael Allen Gillespie, *The Theological Origins of Modernity* (Chicago: University of Chicago Press, 2008), 273, 293.

[23] F. C. Beiser, *The Sovereignty of Reason*, 62.

[24] K. Scholder, *Birth of Modern Critical Theology*, 117.

know things directly. Denial of the facticity of the beginning of Genesis meant that it was no longer determinant for anthropology. Turning the Fall into a myth allows man to assume, with the backing of culture, that he is undimmed and without sin in his rational powers.

It is not science itself which has caused this, for even evolution is a philosophy without convincing evidence. But it has become the religion par excellence for modern man: attributing the cause of material things to chance, not some superior authority. This leaves man alone as the supreme critic. If there is no divine authority over man, man is free to think like a divinity. Since in his body he is clearly not divine, the modern has taken a gift of God, reason, and made it the center of all divine activity.

The Bible has been secularized, so that man may be divinized. While Christ says that the pure in heart will see God, the doubt of all authorities is said to lead to true knowledge, because man sees just fine without the help of a deity (Mt. 5:8). This view of man is "part of the air we breathe in academic theology." "Critical historical study [is] a 'sacred cow' of modern theological study."[25] The consequences are innumerable:

> if no one submission of our natural will or private judgment be demanded of us by *religion;* then not only was the dispensation of the Spirit, and the revelation of Jesus Christ, unnecessary; but *all* revelation was, and is, unnecessary; nay, the existence of one only wise and supreme Governor of all things is unnecessary: for man, thus unbelieving, insubordinate, and independent, is, as it were, a God unto himself!"[26]

The essence of Christianity becomes unscholarly and inappropriate for civilized man well-educated in atheistic methods.

---

[25] W. J. Abraham, *Divine Revelation*, 2.
[26] J. Miller, *The Divine Authority of Holy Scripture Asserted*, 51.

Chapter 17

The Naïveté of Historical Critics

Higher learning is synonymous with critical thinking or *higher* criticism. Man gets above the subject he studies and judges it. Scientific methodology has an air of intellectual respectability for moderns. The scientifically trained scholar is assumed to be without bias or presupposition, which is a massive theological bias. This objectivity, which begins with doubt, is supposed to lead to reliable knowledge. The historical-critical method must be used "unless religious people and communities want a complete breakdown in scholarship."[1] Scholarship, in its approach to knowledge, is the modern salvation.

The new view of authority was combined with the materialistic model of truth to alienate faith from reason and the sensible, tangible world. Modern faith has become so otherworldly it exists only in the mind as a philosophy. When religious teachers talk of method, they do so as naïve children, not realizing that they are signing their religion's suicide note. They should rather preform historical criticism on historical criticism itself, instead of dissecting God's Word. This would be more objective. Sadly, scientism, "the worship of scientific technique," rules in a very narrow-minded way.[2]

---

[1] Raymond E. Brown, *The Critical Meaning of the Bible* (New York: Paulist Press, 1981), viii.
[2] E. L. Long, *Religious Beliefs of American Scientists*, 39.

Divine truth is a bias in natural man's eyes, since it does not come from him. Truth must have equal weight with error, to ensure a fair, critical analysis. Confessional orthodoxy is radically different. It starts with a claim of pure truth, restated in human propositions. Its authority is divine, since the message was received by revelation. But the "historical approach is claimed regarding dogmas or doctrinal articles . . . [so they] need not be taken as eternal truths, but may be set aside as nonbinding, temporally determined matters."[3] Real fundamentalism, defined in 1910, has a limited, paper napkin sort of confession of just five fundamentals. It partakes of the rationalist mindset by limiting the scope of Scripture to just a few controversial matters.

Historical criticism is not unique to Christianity. Its roots are in naturalistic philosophy and it is neutrally objective in that it destroys all authorities. It neutralizes external truth, thereby uniting sinners in their attempt to build a Tower of Babel of atheistic peace and toleration.

> Contemporary Muslims are very interested in developing a hermeneutics of reinterpreting the Qur'an in such a way that the fundamentals of Islam which have universal applicability are separated from historical and cultural accretions which have impeded the growth of Muslim societies, keeping them shackled to a dead past.[4]

The extraction of rational, religious truths from their past context and history is a vain task. Trying to reconcile Cartesian atheism and any historical religion is futile. Historical criticism "by its very nature [is] bound up with criticism of content."[5] Authoritative truth is subjugated and dissolved, so that nothing is left except man-centered toleration and pluralism. "This mass apostasy from God is nothing less than an apocalyptic sign," to those designated as fundamentalists.[6]

The scientific definition of truth is biased against the God of truth. But scholarship of a scientific character is feared, loved, and trusted above all things. In modern thought, to be unscientific is to be untruthful and less than human. The starting point of this "neutrality" is actually

---

[3]C. F. W. Walther, "Answer to the Question: 'Why Should Our Pastors, Teachers, and Professors Subscribe Unconditionally to the Symbolical Writings of our Church? (1858)," in *Walther's Works: Church Fellowship* (St. Louis: Concordia Publishing House, 2015), 20.

[4]Riffat Hassan in *The Fundamentalist Phenomenon*, ed. N. Cohen, 166.

[5]G. Ebeling, *Word and Faith*, 43.

[6]Patrick M. Arnold in *The Fundamentalist Phenomenon*, ed. N. Cohen, 182.

to judge and dole out authority as a divinity.

Can a reading of Scripture be neutral to assumptions about man and knowledge? The Bible claims clearly to be above man, by virtue of the speaking God who inspired it. A strict doctrine of inspiration short-circuits all criticism. It leaves nothing for the scholar to do except believe, rearrange, and apply God's Word. But removing our contact point with God—Scripture—results in "a language-game with Christian terms and an atheistic substance."[7]

Practice, that is, scientific methodology, drives doctrine, when it should be the reverse. A method which distinguishes between words of God, in an accurately reconstructed text of the Bible, allows one to be the master of God's Word. Sinful man sits in methodological judgment over God. Historical-critical methodology "assumes the right to pass rationalistic judgment on Scripture's own claims."[8] To assume that God cannot speak is not neutral for reading God's Word. It is blasphemous and satanic.

The denial of original sin has led the whole academic discipline of exegesis astray. Modern exegesis reformulates the concept of inspiration to make Scripture a scholarly tool.[9] It makes God's Word neutral and non-binding on critical man.

Those who are consistent in historical criticism deny Christ took on flesh and died as the sin-bearer, but to maintain academic respectability, we tolerate what was once called "German mysticism."[10] The right of criticism, the very opposite of obedience and submission, has become the basis for academic theology. This dogmatism of the critics is too restricted by inerrancy of any type, so the very idea has been attacked and disregarded. The critics "were simply inventing a Jesus of their own making, one who uncritically echoed their own ideas, values, and hopes."[11] The true Gospel tastes of death to those who are wrapped in the cords of sin. It is not surprising, by the scriptural standard, that anything but the Gospel is allowed in academic settings. Only

---

[7] Kurt Marquart in *Hermeneutics, Inerrancy, and the Bible*, eds. Earl D. Radmacher and Robert D. Preus (Grand Rapids: Zondervan, 1984), 393.

[8] *Inerrancy*, ed. N. L. Geisler, 84.

[9] D. M. Farkasfalvy, *Inspiration and Interpretation*, 4.

[10] That was the specific charge against this critic in 1882: H. P. Smith, *Inspiration and Inerrancy*, 14.

[11] A. E. McGrath, *Intellectuals Don't Need God*, 151.

the scriptural Gospel can overcome the scholarly institutionalization of atheism.

Theology has voluntarily become a human science. It does not speak for God, but traces the origins of Christianity to sinful men in all sorts of creative ways. For example, Luke supposedly borrowed some themes for writing about God incarnate from the pagan Socrates who committed suicide.[12] Historical criticism offers only human insight of a very limited value. The explanations are historical, but not factual or meaningful. "The notion began to prevail that to have traced out the development of an idea or an institution was to have fully accounted for it."[13] The modern mind is satisfied by so little, which is why pre-modern theology is vastly superior to modern rationalism.

It is true that "Verbal Inspiration has disappeared as if in one night. No theology of any repute now upholds it."[14] Modern theology gets its facts not from God, but atheistic methods, so that "no responsible and eminent theologian today ignores historico-critical research and its questions, dismisses evolution with an 'It is written,' or sets forth a doctrine of revelation without facing historical relativism."[15] A man-centered view of truth leads to a man-centered theology. Method determines results, and therefore historical criticism's appropriateness to deal with God's revelation must be denied.

What is reputable, what will get published and receive acclaim, is rarely of timeless value. An unchanging divine truth is ruled to be unscientific and unscholarly. The true Gospel has become divine parental abuse, a sadistic glorification of death—the most uneducated idea. To accept the unscientific assumption that the Scriptures know more than any reader of them is to be treated as Christ was in His humiliation. To listen in humble repentance to Scripture as Christ's voice is unthinkable for moderns. In the historical-critical understanding of Scripture, the Bible is just one of many sources for "respectable professional exegesis."[16] But even its most strident defenders recognize

---

[12]Peter J. Scaer, *The Lukan Passion and the Praiseworthy Death* (Sheffield: Sheffield Phoenix Press, 2005).

[13]H. D. McDonald, *Theories of Revelation*, 2:80.

[14]Reinhold Seeberg, quoted in: T. Engelder, "Verbal Inspiration: A Stumbling Block," Part 11, 233.

[15]H. Thielicke, *Prolegomena*, 36.

[16]Karlfried Froelich in *Studies in Lutheran Hermeneutics*, ed. J. Reumann, 136.

the massive danger it poses to the substance of Christianity: historical criticism is "the most serious test that the Church has had to face through nineteen centuries."[17]

---

[17]E. C. Blackman, quoted in: E. Krentz, *The Historical-Critical Method*, 4.

Part VI
Judging or Being Judged

Chapter 18

The Sin of Hermeneutics

How is understanding possible? The aim of modern scientific hermeneutics is to answer this question. Men who live and move and have their being in God attempt to explain the science of knowledge without the hypothesis of God. No wonder modern hermeneutics is a complete dead end, resulting in nothing but man's authorizing of pure and unbridled subjectivity. Atheistic talk of understanding leads to none at all. Hermeneutics, as the reigning philosophy of modernism, can say nothing certain, except that language itself is uncertain. It starts with the scientific observation of the "inadequacies and brokenness of all human attempts to state the truth," despite the fact that communication is frequently made between humans in practical affairs.[1]

Man-centered attempts to objectively ground human language and understanding have failed, resulting in the judgment that "there is no such thing as purely objective interpretation."[2] While it sounds like a more Christian approach, it is a critical judgment. Modern interpretation of Scripture, no matter its results or disclaimers, starts with man, not its Author. A presupposition of "modern exegesis" is "the recognition of the modern worldview," including the divinity of man's mental powers.[3] Though it announces the failure of objective historical criticism, the postcritical approach restricts man no more than classical Enlightenment philosophy. Instead of declaring the man-centered view of truth to be an

---
[1] David Tiede in *Studies in Lutheran Hermeneutics,* ed. J. Reumann, 283.
[2] P. H. Nafzger, *These Are Written*, 147.
[3] K. Scholder, *Birth of Modern Critical Theology*, 132.

utter failure and repenting, it posits that all truth is beyond formulation in human words. As an evidence-based argument it is eminently modern and does not wish to combat any supposed "scientific facts." Man, specifically the specialist scholar, replaces the Holy Spirit in the modern interpretive framework.

"Hermeneutics presupposes the problem of historical distance and the rise of historical consciousness."[4] The Word of God is essentially inaccessible and incompatible with man in modern hermeneutics. Its subject matter is instead the rational act of understanding.[5] It has little to do with Scripture. The word "inspiration" may be lauded, but its import does not affect modern hermeneutics. The Bible is treated as a human document to be scientifically studied.

A defining characteristic of modern exegesis is that no matter what confession of the inspiration and authority of Scripture is made, it has little bearing on exegetical practice. In the first 15 centuries of the Church, "we can see a close interdependence between theories of inspiration and hermeneutical practices."[6] Modern hermeneutics, on the other hand, treats Scripture scientifically as any other writing, not with a special method.

For moderns, "hermeneutics [is] the very essence of philosophy."[7] Even for biblical scholars, scriptural interpretation is a subset of the secular quest to understand understanding scientifically. Modern theologians, mostly university professors, "tried more or less deliberately to build a bridge between the gospel and their secularized cultural environment."[8] The critic claims that being objective and without presuppositions is how to read the Bible. "Since when is color blindness necessary for the history of painting, or tone deafness for the science of music?"[9]

While moderns claim that the Spirit is not needed, scholarly skills are essential: exegesis is "technically impossible to do" except in the

---

[4]Anthony C. Thisleton, *The Two Horizons: New Testament Hermeneutics and Philosophical Description* (Grand Rapids: Eerdmans, 1992), 51.
[5]G. Maier, *Biblical Hermeneutics*, 15.
[6]D. M. Farkasfalvy, *Inspiration and Interpretation*, 12.
[7]G. Goldsworthy, *Gospel-Centered Hermeneutics*, 134.
[8]Hendrikus Berkhof, *Two Hundred Years of Theology: Report of a Personal Journey,* trans. John Vriend (Grand Rapids: Eerdmans, 1989), xiii.
[9]Albrecht Oepke, quoted in: G. Maier, *Biblical Hermeneutics*, 44.

original languages.[10] As desirable and praiseworthy as the languages are, possessing God's Word is not contingent on technical skills. A person with a lousy Bible translation may receive blessed faith in Christ and grasp the mystery of the Gospel. The scholar who approaches the Scriptures with all the necessary critical skills may be treating it as malarkey and hate Christ critically. Scripture is not an object to judge. Plenty of people speaking Jesus' native language and idiom heard Him and rejected Him. The Holy Spirit is the only necessary presupposition for understanding God's Word.

The Small Catechism defines the work of the Spirit as creating faith in the sinner and illuminating him. Confessing Christ is also a confession that the Spirit "enlightened me with His gifts." This inner "illumination of the minds of all believers by the Holy spirit makes no new revelation of truth but uses for his instrument the truth already revealed."[11] While the Spirit is the only presupposition for understanding God's Word, it is not something man brings to the text. Luther in his Genesis Commentary relates that "the Holy Spirit has his own language and way of expression," including "his own [technical] terms."[12] The Spirit comes through the words, which are His, and reveals Christ to the sinner to save him.

A sinner can only bring unbelief to the Word of God. "Whatever reason brings out of its stinking heaps must be subordinate to the Word of God . . . ."[13] The Spirit—properly and fully uncreated God—comes through the truth of God's Word.[14] The Word itself brings the Spirit, meaning no presuppositions of understanding or rational assumptions make understanding God's Word possible. Of course, the Bible is in plain human language, not in unspeakable, spiritual words or "things that cannot be told, which man may not utter" (2 Cor. 12:4). The Word of Christ can, and does, go out into all the world. The Spirit

---

[10] David L. Adams in *The Bible in the History of the Lutheran Church*, ed. J. Maxfield, 37.

[11] Augustus Hopkins Strong, *Systematic Theology: A Compendium and Commonplace-Book Designed for the Use of Theological Students* (New York: A. C. Armstrong and Son; 2nd ed., 1889), 17.

[12] LW 1:47.

[13] Nicolaus Arnoldus, quoted in: K. Scholder, *Birth of Modern Critical Theology*, 119.

[14] "In, with, and under" is apt language for the relationship of the Spirit to the Word.

makes believers through the Word, and without Him there is no faith or understanding of Christ. The problem with apprehending the Christian revelation is not a flaw in man's rational faculty, it is unbelief and rebellion against the voice of the Lord's Anointed.

Who is the fullest and highest author of Scripture? As a computer warning states, knowing where something ultimately came from is essential to dealing with it appropriately: "only install this file if you trust the origin." That is why the Spirit's inspiration was deemed worthy of being included in the Nicene Creed—if Scripture's origin is God the Spirit, then the words can be trusted. If it is the Spirit's book, then interpreting Scripture is not like measuring gravity or weighing objects. The Holy Spirit is not an object we can assess by naturalistic means, since He is the cause of all things.

God has sets limits to what man can know. Science must admit it does not have absolute truth, since it is always critiquing its own results and evolving in its understanding. That does not mean we must redefine truth, but that truth based on man is doomed to failure. The scientific method treats man as God, but he is not, and most things are above him, including the science of understanding everything: hermeneutics.

The hermeneutical circle is bandied about as the paradigm for understanding, because man is *a priori* assumed to be on the same level as God. In the circle, they are of equal standing. Indeed, God cannot speak or act without man's verification, so modern hermeneutics is by definition only about man. In scientific fashion, what is outside of man is ruled out of bounds and not scientific. This is the wisdom of philosophical hermeneutics: it can only cloud God's Word and make everything unintelligible. "Has not God made foolish the wisdom of the world?" (1 Cor. 1:20). The idea of the hermeneutical circle shows that man is enclosed in a circle of sin and under God's judgment. Thoughts which originate in man lead not to truth, but to death.

"The ultimate aim for hermeneutics is to understand the author better than he understood himself."[15] The original author's authority is marginalized, thereby "interpreting the reader is as important as interpreting the author."[16] While it is dressed with scholarly language,

---

[15] Wilhelm Dilthey, quoted in: G. Maier, *Biblical Hermeneutics*, 17.

[16] John Breck, "Exegesis and Interpretation: Orthodox Reflections on the 'Hermeneutic Problem'," *St. Vladimir's Theological Quarterly* 27:2 (1983), 91.

the core of hermeneutics is man who authorizes God's Word. God cannot speak directly to man, since His words cannot be excluded from human conditioning and accommodation to error. The science of interpretation is a dead end, if "we listen for a Word from God in which the text that once was God's Word to people in a different time and place might become God's Word for us."[17]

Hermeneutics has become the works righteousness of the modern era—a righteousness that appears better than Christ's justification. Man undertakes with great effort to make God relevant to a rebellious, faithless people. Modern idolatrous works are not outward, as in biblical times, but primarily mental and intellectual. The shift from physical idolatry to mental idolatry was profoundly described in 1817:

> The apostasy of the Jews became IDOLATRY; a gross, palpable crime: the apostasy of modern times appears to be a SPIRITUAL and INTELLECTUAL REJECTION of the Deity; either wholly, or at least in part, as now predicated in his mysterious essence. A portentous form of infidelity! resulting from the abuse of "liberty" into "licentiousness;" from the pruriency of that more subtle part of the constitution of human nature, to which the Gospel addresses its appeal, uninfluenced and unrestrained by that fundamental submission of the will, which it inculcates and insists upon.[18]

Submission to anything has become the idol for moderns. The thoughts of man have become the only acceptable gods.

Of course, objectivity in the sense of being a god is not possible for man. How can one pin down the Spirit for study when he cannot even leash air currents and weather patterns (Jn. 3:8)? Yet, in becoming an object for scientific experimentation, the Bible's revelatory quality is quietly denied. "The abstract notion of a 'revelation' is now difficult even to be conceived; because the revelation of the Bible, which has so grown up with us and insinuated its influence throughout all our faculties, has so completely undeceived us."[19] Because moderns have rebelled and are distant from Christ, the Bible must be hermeneutically translated for atheistic man.

"Hermeneutics makes the Bible relevant to modern man," so that God is unable to speak to grown-up men today, or rather, men are

---

[17] Duane Priebe in *Studies in Lutheran Hermeneutics*, ed. J. Reumann, 296.
[18] J. Miller, *The Divine Authority of Holy Scripture Asserted*, 46–47.
[19] J. Miller, *The Divine Authority of Holy Scripture Asserted*, 96.

closing their eyes and rebelling against the truth of Christ.[20] In contrast, in pre-modern hermeneutics the truth is always connected to the Spirit. Reading and teaching Scripture is a spiritual endeavor, since the highest act of worship is to hear and speak God's Word.

> For of all acts of worship that is the greatest, most holy, most necessary, and highest, which God has required as the highest in the First and the Second Commandment, namely, to preach the Word of God. . . . If this worship is omitted, how can there be knowledge of God, the doctrine of Christ, or the Gospel?[21]

---

[20] John Breck, "Exegesis and Interpretation," 77.
[21] *Apology of the Augsburg Confession* (AP) XV:42.

Chapter 19

Sacred Hermeneutics

No hermeneutic which ignores sin can be authentically Christian.[1] *Sola Scriptura* is the highest absurdity to deified man, because it makes him a dependent. Yet Scripture itself cries for a special, sacred hermeneutic.[2] Its own claims mean that either it is an evil fraud or it must be treated totally unlike all man-made writings. Interpretation is an art, not a science. Even more, it is ultimately a spiritual exercise to comprehend God's true Word in faith. "Only regenerate people can truly understand divine truths."[3]

What the modernist person cannot grasp is God's involvement in this world and in His Word: "Fundamentalism is a mode of thought which, seemingly basing itself on the doctrine of verbal inspiration, disregards the diversity of Scripture and the hermeneutical problem in general."[4] The counterpart to a revelation in history is the Spirit's present involvement. The problem with modern man is the denial of sin which prevents the apprehension of divine teaching. Neutrality before God, or assuming that the Spirit is not needed to read Scripture, is blasphemy. The problem in dealing with Scripture is not hermeneutics in general, but unbelieving hearts and minds which can only hate God without the Spirit. "For human hearts without the Holy Ghost are without the fear of God; without trust toward God, they do not believe

---

[1] G. Goldsworthy, *Gospel-Centered Hermeneutics*, 188.
[2] G. Maier, *Biblical Hermeneutics*, 21.
[3] O. Valen-Sendstad, *The Word That Can Never Die*, 9.
[4] H. Berkhof, *Two Hundred Years of Theology*, 132.

that they are heard, forgiven, helped, and preserved by God. Therefore they are godless."[5]

The bedrock of modern hermeneutics is historical distance, that is, God's total abstention from worldly contact. Yet, orthodoxy sees God's involvement not as special and remarkable, but regular and the backbone of all true knowledge. The Spirit overcomes all historical uncertainties and hermeneutical problems by His own witness. But this supernatural explanation is not empirically observable, as Jesus expressed to a rationalist: "The wind blows where it wishes, and you hear its sound, but you do not know where it comes from or where it goes. So it is with everyone who is born of the Spirit" (Jn. 3:8).

There is much to ponder in the assertion that the first of the Ten Commandments—having no other gods—is "the primary hermeneutic framework."[6] Yes, the law of God even reaches the scholar on his critical throne. Handling the Word of truth is a spiritual endeavor, because in the Bible the Lord of all creation deals with sinners. One must not approach as the lord of the text, eager to teach God, but listen in humble submission. "To disclaim technical mastery of a text is in fact one important hermeneutical principle: the principle that one must always listen."[7] One must follow the words and their teachings to be taught by God: "'And they will all be taught by God.' Everyone who has heard and learned from the Father comes to me—not that anyone has seen the Father except he who is from God; he has seen the Father" (Jn. 6:45–46). To believe is a "sacrifice of the (sinful) intellect," but the gain is life eternal. Critical, know-it-all judgment is included in "all things": "Indeed I also count all things loss for the excellence of the knowledge of Christ Jesus my Lord" (Phil. 3:8).

Even in a secular art, like painting, excellence requires more than knowing rules or what non-artists say about the subject. It requires not scientific objectivity, but some spark which the non-artistic do not have. In theology the Spirit is necessary to think and speak correctly about Christ. "No one can say, write, or think anything beneficial, comforting, or noteworthy about Jesus without God's Spirit."[8] It is

---

[5] AC XVIII:72.
[6] G. Goldsworthy, *Gospel-Centered Hermeneutics*, 74.
[7] B. A. Gerrish in *The Seven-headed Luther,* ed. P. N. Brooks, 15.
[8] Valerius Herberger, *The Great Works of God: Parts One and Two: The Mysteries*

correct to speak of a dual inspiration, as long as only Scripture is the revelation of information. The Scripture's inspiration is the foundation of hermeneutics and exegesis. Without it, the Spirit is cast out and the Bible becomes naturalized—a human object of scientific study.

Communication does not start with a receiver but with an author. God's revelation invalidates all assumptions based on human language or understanding. Luther states: "no one understands the spiritual Scripture unless he tastes and possesses the same Spirit" who wrote it.[9] Luther did not invent this idea; it was the pre-modern Christian approach to truth: St. Bernard of Clairvaux, 400 years prior to Luther, said the same: "the Holy Ghost inspireth the true meaning of the Scripture to us: in truth, we cannot without it attain true saving knowledge."[10]

Understanding the Bible's origin is crucial to understanding its contents. Scripture was "produced under divine influence, so also it is to be understood according to enlightenment by the same Spirit." This was generally held "by all Church Fathers and all medieval theologians."[11] This is exactly how Luther explains the third article of the Creed: "*I believe* that by my own reason or strength that *I cannot believe* in Jesus Christ my Lord or come to Him." The Spirit gathers, enlightens, and sanctifies incompetent and unbelieving man. "God, not the official teachers of the church, must interpret his word," and without the Spirit it is not understood or communicated.[12] This is the basic teaching of God's Word:

> Now we have received not the spirit of the world, but the Spirit who is from God, that we might understand the things freely given us by God. And we impart this in words not taught by human wisdom but taught by the Spirit, interpreting spiritual truths to those who are spiritual (1 Cor. 2:12–13).

The Word of God cannot be received by natural man. Mountains of man-centered knowledge are useless for capturing divine truth. Sin

---

of Christ in *The Book of Genesis, Chapters 1–15,* trans. Matthew Carver (St. Louis: Concordia Publishing House, 2010), 15.

[9] Quoted in: G. Maier, *Biblical Hermeneutics,* 48.

[10] T. H. Horne, *An Introduction to the Critical Study and Knowledge of the Holy Scriptures,* 491.

[11] D. M. Farkasfalvy, *Inspiration and Interpretation,* 57.

[12] G. Goldsworthy, *Gospel-Centered Hermeneutics,* 195.

is not erased by education; rather, it paves the way for a more subtle idolatry. Reason, by itself, is incapable of grasping God's Word. "So the blemish of obscurity is not in the divinely inspired scriptures but in the minds of men, who walk in the vanity of their senses, having an intellect obscured with darkness, alienated from the life of God by the ignorance that is in them, . . . [or blinded by] the fraud and power of the devil."[13] Death to critical, god-like aspirations is part of the Christian life. "He himself bore our sins in his body on the tree, that we might die to sin and live to righteousness" (1 Pet. 2:24). Death to sin includes death to wanting to know more than God reveals. Hugh of St. Victor warns: "Do not despise the humility of the Word."[14] Jesus says to the erudite modern: "Yes; have you never read, 'Out of the mouth of infants and nursing babies you have prepared praise'?" (Mt. 21:16).

Scripture is its own key. However, this does not necessitate a flat reading, or that every word is ripped out of its inspired context and applied indiscriminately. *Sola Scriptura* is not a rational framework. It means that the Spirit in the words should be listened to above all human thoughts. "The Holy Spirit is at same time the author and the interpreter of Scripture."[15] The Spirit provides His own context and does His own teaching, though sinners can only reject the divine Word according to their flesh. The problem is not Scripture's original historical context—being a Jew at the time of Christ, or even in His own family, did not guarantee spiritual understanding. There is not a historical distance between the Scriptures and the reader, but a spiritual one.

Divine inspiration must frame all hermeneutical and interpretive questions, or atheistic methods will. God gave the words the human authors wrote. That is the only way the Bible can truly be considered "God's Word" in any literal sense. This does not mean that everything recorded in it is sanctioned. Inspiration does not speak to the applicability of a particular passage. Man does not predetermine how God must speak and come to man. Rather, he can only accept in the Spirit or reject in unbelief what is revealed. The problem of scriptural hermeneutics is

---

[13] Johann Major (d. 1654), quoted in: K. Scholder, *Birth of Modern Critical Theology*, 148.

[14] G. Goldsworthy, *Gospel-Centered Hermeneutics*, 105.

[15] Matthias Flacius, *How to Understand the Sacred Scriptures,* trans. Wade R. Johnston (Saginaw, MI: Magdeburg Press, 2011), 67.

nothing like the interpretation of writings originating in man: "The interpretation of legal documents is greatly akin to scriptural interpretation, and many of the issues currently confronting Biblical interpreters confront interpreters of, e.g., the United States Constitution."[16] Though if exegetes cannot say anything definitive about purely human documents, their ability to deal with divine Scripture is extremely suspect.

The unity of modern, rational man is assumed above the unity of God's own speech. Hermeneutics has the appearance of being scientific, but by all classic definitions God is above man, not an object to be thrust under a microscope. It is in reality an unscientific rationalism that is applied to all things divine. Scripture is either to be rejected or accepted—but only one approach can be genuinely Christian. "The reasonable critic must not be closed to the possibility of disunity" in Scripture.[17] But for an "open examination of Scripture," the "whole dogmatic tradition [is] one great prejudice."[18] This "all or nothing" position seems extreme, but as Luther describes David's claim to inspiration, an oracle requires it: "What a glorious and arrogant arrogance it is for anyone to dare to boast that the Spirit of the Lord speaks through him and that his tongue is the Word of the Holy Spirit! He must obviously be sure of His ground."[19] In reading Scripture, it is divine speech that we read: "to me it seems to be the height of absurdity if we think that the Holy Spirit immediately broke off His speech."[20]

It is very reasonable that Scripture may be false, but unbelief and starting with a denial of the scriptural message is not an eminently Christian activity. Modern hermeneutics starts with the assumption that man is clearer than God. The pre-modern view was not always theologically correct, though at least it recognized the need for divine enlightenment. Luther, who had one of the strongest doctrines of man's sin since the New Testament era, had anthropological reasons to extol revelation.[21] He confessed, as an article of faith, that which is only

---

[16] James W. Voelz, *What Does This Mean?: Principles of Biblical Interpretation in the Post-Modern World* (St. Louis: Concordia Publishing House, 1995; 2nd rev. ed., 1997), 13.

[17] John Piper, "A Reply to Gerhard Maier: A Review Article," *Journal of the Evangelical Theological Society*, 22:1 (1979), 85.

[18] K. Scholder, *Birth of Modern Critical Theology*, 134, 137.

[19] Luther, *On the Last Words of David* (1543), LW 15:275.

[20] Commentary on Gen. 49:11–12, LW 8:247.

[21] Luther's *The Bondage of the Will* (1525) is not about God's absolute, mechanical

possible in the Spirit: "Be it known, then, that Scripture without any gloss is the sun and sole light from which all teachers receive their light, and not the contrary."²²

---

ordaining, à la the Calvinists, but man's spiritual incapability—even from his creation. It certainly has not changed for the better since the Fall. "Here we are speaking not only of the first man, but of any and every man, though it is of little importance whether you understand it of the first man or of other men, for although the first man was not impotent when he had the assistance of grace, yet by means of this precept [to not eat of the fruit] God shows [Adam] plainly enough how impotent he would be in the absence of grace. But if that man, even when the Spirit was present, was not able with a new will to will a good newly proposed to him (that is obedience), because the Spirit did not add it to him, what should we be able to do without the Spirit in respect of a good that we have lost?" LW 33:124.

²²Quoted in: J. M. Reu, *Luther and the Scriptures*, 77–78.

Chapter 20

Unity of Attitude toward Scripture and Christ

Christ the God-man is not currently open to examination in His exalted state. The Bible, a divine writing of miraculous origin, is open to man's scrutiny and investigation. What we study is the result of the Spirit's breathing and where God allows Himself to be handled. Though Scripture is abused and maligned as erring and weak, that does not mean it is less than divine and truthful.

Human methods are powerless in the face of God's eternal Word. There is no decisive method or list of steps to follow for interpretation. The necessary ingredient is God's Spirit who gave the spiritual words. The Spirit will not be cajoled by any method, especially one that allows man to think without limits as a god.

Many are weary of Scripture talk, and with good reason. We are not saved by Scripture's inspiration. However, since Scripture is the basis for theology, it cannot be given up to the dogs of rationalism. "With the reality of revelation, therefore, Christianity stands or falls."[1] How one treats Scripture is indicative of how one views Christ: "For whoever is ashamed of me and of my words in this adulterous and sinful generation, of him will the Son of Man also be ashamed when he comes in the glory of his Father with the holy angels" (Mk. 8:38). Christ identifies Himself

---

[1] Herman Bavinck (d. 1921), quoted in: B. Ramm, *Special Revelation and the Word of God*, 16.

with His words. We only know of Christ in words right now.

Meager as the Bible appears, it is still in demand. Though historically distanced, it is still a best-selling book. The charge to scholars and critics is a concrete one, which requires more than haughty criticism: "Ho, all ye scholars and critics! do but write such a book, and we will believe you!"[2] For all the talk of Christ, there is no authority of Christ apart from the authority of Scripture.

Luther did not downplay the apparent warts of Scripture: like Christ, the Bible is "a worm and no book, compared with other books."[3] God has hidden Himself in the words. The "interchangeability of the terms *God* and *scripture* in certain New Testament passages" indicates that Scripture is equivalent to the divine will.[4] To know Scripture rightly is to know God's will. Jesus' charge to critics shows that to be ignorant of one is to be ignorant of both: "you know neither the Scriptures nor the power of God" (Mk. 12:24). Historical-critical methods, however, start with the idea that there is something to critique in Scripture. The "reader is set over Holy Scripture as judge, that he should distinguish what is essential and nonessential, what is divine and not divine, what is true and false."[5]

"For the Church and her members, there is no need for proof for the inspiration of Scripture, for her very existence depends upon this faith, and this faith precedes all proofs."[6] The Church is not founded by faith in the idea of Scripture, but the words and power of the Spirit. The pure Word of God previously came in various forms, but we only find the Spirit for us in the words of Scripture, as a fountain: "We pledge ourselves to the prophetic and apostolic writings of the Old and New Testaments as the pure and clear fountain of Israel, which is the only true norm according to which all teachers and teachings are to be judged."[7] The basis for the Church and the individual's acceptance of Scripture is the Spirit, who is in His words. A human basis would make it of merely human authority and conviction. The testimony of the Spirit is that important teaching that God Himself—not man—authorizes His

---

[2] F. Bettex, *The Bible the Word of God*, 52.
[3] K. E. Marquart, *Anatomy of an Explosion*, 44.
[4] John W. Wenham in *Inerrancy*, ed. N. L. Geisler, 21.
[5] C. F. W. Walther, "Evening Lectures on Inspiration," Lecture I.
[6] H. Schmid, *The Doctrinal Theology of the Evangelical Lutheran Church*, 51.
[7] FC SD Rule, 3.

Word.

> By the internal testimony of the Holy Spirit, is here understood the supernatural act of the Holy Spirit through the Word of God, attentively read or heard (His own divine power being communicated to the Holy Scriptures), moving, opening, illuminating the heart of man, and inciting it to obedience unto the faith.[8]

To demand proof or evidence for God's speech is to deny its divine character.

While Scripture is not absolutely necessary, the external Word of God is certainly essential for salvation. "Christ in us is not more privileged than Christ outside of us."[9] Scripture is God's Word—the only pure access to Christ. God the Spirit, not man, verifies and authenticates that Word in every believer. "Only the Word [of God] creates divine assurance" and true spiritual unity.[10] Because Scripture is God's speech, inerrancy is a basic expectation, but also one established by Scripture. The possibility of even one error invalidates the whole thing as being from God. In that case, God would no longer be the supreme authority over the Christian. If there is even the potential for error in Scripture, the reader is actually the supreme judge and norm of truth. Either Christ speaks in Scripture or the text becomes "a mouthpiece for [the critic's] own prophetic utterance."[11]

Christ is a speaking person: "I still have many things to say to you, but you cannot bear them now" (Jn. 16:12). His words carry the authority of His person, which overrules any sovereignty established by method. Jesus even refers to the individual letters of Scripture: "For truly, I say to you, until heaven and earth pass away, not an iota, not a dot, will pass from the Law until all is accomplished" (Mt. 5:18). "The biblical Christ was concerned with the inerrancy of jots and tittles," that is, the smallest parts of written letters.[12] This is deemed unworthy of Christ by divinized man, so Christ's own method is dismissed as fundamentalistic without even an apology. That so many moderns do not

---

[8] David Hollaz (d. 1713), quoted in: H. Schmid, *The Doctrinal Theology of the Evangelical Lutheran Church*, 55–56.

[9] H. D. McDonald, *Theories of Revelation*, 1:203.

[10] C. F. W. Walther, quoted in: Theodore Engelder, *Haec Dixit Dominus: Thus Saith the Lord* (St. Louis: Concordia Publishing House, 1947), 6.

[11] Albrecht Oepke, quoted in: G. Maier, *Biblical Hermeneutics*, 322.

[12] J. W. Montgomery, *Crisis in Lutheran Theology*, 37.

even try to justify their methods and starting principles from Scripture is telling. *Sola Scriptura* is a medieval relic to them.

While Luther is cited by all parties in the modernist debate, many of his statements are irreconcilable with modernism. He states emphatically: "one letter, even a single tittle of Scripture means more to us than heaven and earth."[13] Clearly Luther did not stop with the letters, but through them he found Christ. The letters are greater than heaven and earth, because they are eternal. "The smallest letter will not pass away, neither a jot nor a tittle (little hook). With that [Christ] wants to say: 'The smallest letter will not pass away, not even half a letter'."[14] The eternally begotten Word, Christ, does not speak temporary words through the Spirit. The very letters bring us Christ and cannot pass away.

Scripture contains enough divine truth so that the scholar can drown, but the elementary student may be edified. Gregory the Great rightly said: "Scripture is like a river . . . broad and deep, shallow enough here for the lamb to go wading, but deep enough there for the elephant to swim."[15] The Bible has a specific purpose. Its words and truth serve the purpose of salvation in Jesus. Christ did not die so that we could play patty-cake with His words. There is truth to be believed in divine certainty and proclaimed to set sinners free in Christ. The true Christ is known only in His words, which are found only in the Scriptures. Any other Christ is silent and not the true Savior.

---

[13] Quoted in: J. M. Reu, *Luther and the Scriptures*, 83.
[14] C. F. W. Walther, "Evening Lectures on Inspiration," Lecture VI.
[15] *Commentary on the Book of Blessed Job*, Epistle, 4.

Chapter 21

Textual Criticism

The oral speech of the apostles was the standard for the first generation of Christians. Speech is a difficult norm to sustain over time. The fact of the written Word of God is problematic today, precisely because it is human and observable. It is open to every critique and atheistic method. It seems to just silently submit. But man as the ultimate criterion of truth is utterly individualistic. Despite its own claims, post-modernism shows the true end of modernism: everyone constructs his own truth and reality.

"Reason has a way of turning upon itself."[1] Post-critical thought is nothing more than an affect, a reveling in the uncertainty of words, while avoiding the contentious issue of truth. The return to external authorities, such as the clear words of the Bible, is not on the agenda of the hyper-rational post-modernists. One of the main attacks today, besides the denial of the communicative property of language, is the attack on the physical text of Scripture.

A written text is not a problem for the believer, but a divine gift. Written language is the most stable form of communication. It is true that we do not have the original manuscripts of Scripture, but we have more than sufficient warrant to be sure of their divine teaching. It is not objectivity or ruling out all academic questions that establishes faith, but God's own Spirit in the believer. "What the Church lacks in our day is not a reliable text of the Bible, but faith in the sufficiently

---

[1] J. C. Livingston, *Modern Christian Thought*, 40.

reliable text."[2] For all the talk of variants, there is a general failure to point to where a variant specifically undermines a Christian doctrine. Instead, a minor uncertainty in a particular manuscript out of thousands of manuscripts is used to logically destabilize the whole structure of Scripture. However, divine knowledge is not based on completeness in the scientific sense; it is revealed by the Spirit in words. "There are very few questions of real significance about the actual wording of either testament," but critics want to find problems and play the part of the theologian-philosopher, lording themselves over God's clear words.[3] This real issue is what grounds knowledge and grants certainty: God's words or man's.

Assuming there is no original text is a denial of God's Word. "It is apparent that scholars increasingly question whether there is only one original text for a given [biblical] book."[4] Inerrancy only applies to the originals, which infuriates critics because it does not give room to judge the hypothesis of inerrancy. This qualification that inspiration applies only to the autographs is necessary, despite the fact that man cannot pass judgment on them. If I tear a page of out my personal Bible, is that God the Spirit's fault? Does it change what St. Paul wrote down 2,000 years ago? Inspiration is not an ongoing process, it has to do with the historical writing of particular words. We can safely assume that the existing words in our Bibles had a beginning. Only by faith in His words can we trust that the Spirit is the origin of them. All arguments against Scripture based on distance from the originals is historical criticism. Man's decision, even by the smartest scholars, cannot validate anything as God's Word. The originals would not create more faith than our copies of copies, nor would they satisfy insatiable critics.

It is a wrong-headed assertion to say that textual criticism is new. "Beginning with Paul Kretzmann [d. 1965], the distinction between 'inspired autographs' and copies is presented in the language of Princeton and Fundamentalism" in the LCMS.[5] How a supposed textual expert

---

[2] Francis Pieper, *Christian Dogmatics,* 4 vols., trans. T. Engelder, J. T. Mueller and W. W. F. Albrecht (St. Louis: Concordia Publishing House, 1953), 1:340.

[3] Allan A. MacRae in *Hermeneutics, Inerrancy, and the Bible,* eds. E. D. Radmacher and R. D. Preus, 153.

[4] J. W. Voelz, *What Does This Mean?,* 80.

[5] Jeffery Kloha, "Kloha's Response to Montgomery Essay," *Christian News* 52:4 (Monday, Jan. 27, 2014), 14.

can claim this is a novel, Reformed distinction, contrary to the plain record, is unfathomable. The distinction has always been made by good Christian thinkers. In 1893 the Roman papacy declared: "It is true, no doubt, that copyists have made mistakes in the text of the Bible."[6] The qualm is not with "autographs," but "inspired" originals, which would not be open to criticism, even if they did still exist. Facts which cannot be verified by man are to be doubted by the Cartesian method.

Contrary to modern know-it-alls, Luther posits errors in manuscript copies, but not the original writings inspired by the Spirit. On the disputed section of 1 Jn. 5:7-8, commonly called the *Comma Johanneum*, the Reformers were not afraid to critique the manuscripts they possessed. Luther said that the suspect verses, which the "Greek codices do not have," were "added by some ignoramus." Luther's compatriot John Bugenhagen called the addition an "Arian blasphemy" and argued strongly for its deletion.[7]

On several occasions Luther was not afraid to criticize a human mistake, but never God's Word. He commented on Acts 13:20: "It is the error of a scribe, who wrote four instead of three, which could easily happen in Greek." Regarding the names recorded in 1 Kings 5:19, Luther writes: "Here we believe that the [Hebrew] text has been corrupted by a scribe, probably because of a copy in illegible writing and through poor letters."[8]

While textual criticism wants to be higher criticism, real textual criticism is a small, thankless, and drudgerous task. It does not allow one to make up bold, creative claims, unless one makes the whole Bible uncertain and full of multiplicities. Textual criticism is real scientific work based on facts. It is not exciting, revolutionary, or prophetic work—unless one is bad at it. Revelation "used normal human language," so we need to be able to identify the words.[9]

A staunch supporter of inspiration is not disturbed by the apparent humanity of the words. The Spirit did not have to reveal all the things recorded by the human writers, but He chose to speak through them for

---

[6] Pope Leo XIII, "*Providentissimus Deus*: On the Doctrine of the Modernists," 20.

[7] Franz Posset, "John Bugenhagen and the *Comma Johanneum*" *CTQ* 49:4 (1995), 246, 248.

[8] Quoted in: J. M. Reu, *Luther and the Scriptures*, 58.

[9] G. Maier, *Biblical Hermeneutics*, 75.

our certainty. What is important is that the Bible we use is dependable, that is, God's absolute Word. Prior sources to Scripture and the personal history of the writers do not argue against the Spirit. Rather, the Spirit used believers and those who had a vested interest in the contents of Scripture. Why is the Holy Spirit not allowed more freedom than the human interpreter?

The New Testament cites translation in its use of the Old Testament. This tells us that the Word of God is easily translated without losing its authority. This propositional quality makes it possible to call a good translation fully God's Word. In the words of a recent confession: "We further affirm that copies and translations of Scripture are the Word of God to the extent that they faithfully represent the original."[10] This was also the position of the scriptural Christ. "Jesus referred to copies handed down of the Old Testament," not Moses' original autograph, yet He still calls it God's speech:[11]

> And Pharisees came up to him and tested [Jesus] by asking, "Is it lawful to divorce one's wife for any cause?" He answered, "Have you not read that he who created them from the beginning made them male and female, and [the Creator] *said*, 'Therefore a man shall leave his father and his mother and hold fast to his wife, and the two shall become one flesh' " (Mt. 19:3–5)?

"According to Jesus, the voice of the world's Creator is the same voice that speaks through the human words of the Book of Genesis."[12] If Christ be true God, His knowledge of Scripture is also true and accurate.

Are difficulties in the text to be attributed to God or man? Whereas accuracy and truth have a scientific pale today, God's truth is not concerned with exact precision. That there is a "such a freedom and flexibility in the mode of Old Testament citation should warn us of the mystery of inspiration and prevent us from claiming to know too much of its process."[13] This does not mean that we can make ourselves

---

[10] Chicago Statement (1978), quoted in: Norris C. Grubbs and Curtis Scott Drumm, "What Does Theology Have to Do with the Bible? A Call for the Expansion of the Doctrine of Inspiration," *Journal of Evangelical Theological Society* 53:1 (March 2010), 68.

[11] C. H. Pinnock, *A Defense of Biblical Infallibility*, 16.

[12] Scott W. Hahn, "For the Sake of our Salvation: The Truth and Humility of God's Word," *Letter & Spirit* 6 (2010), 23.

[13] B. Ramm, *Special Revelation and the Word of God*, 56.

equal to the Spirit, but we should recognize that the Spirit does not care about our conventions of footnotes and scholarly accuracy. Truth is not exactness; rather, it relates a person, Christ, the Lamb of God.

To say "no one knows what was in the originals" is to submit God to the scientific canons of man. Scholars can dismantle Shakespeare as well as Jesus, but they can offer no solid proof or certainty. We can trust the Bible, which was taken seriously as a divine book before the printing press.

"Critics of inerrancy argue as if the original manuscripts never existed."[14] The issue is not whether we can verify the originals, but whether God has spoken to justify the godless. If the words are from God, we have to take them seriously, which means getting as close to the original words as possible. If the historical act of inspiration is denied, the textual task will be fraught will bad assumptions and prejudices. The result can only be human words, which are not taken seriously. There are plenty of clear words to digest and treasure, so this lower textual criticism is not the final step, but merely the initial one. As it turns out, an almost perfect text is sufficient for completely imperfect readers to establish perfect doctrine.

---

[14]Bernard Ramm, *Protestant Biblical Interpretation: A Textbook of Hermeneutics for Conservative Protestants* (Grand Rapids: Baker Books; 3rd rev. ed., 1999), 208.

Chapter 22

The Canon

The canon has received much attention recently. There is even a new criticism based on it called "canonical criticism." This criticism is organized around an assumed or functional unity, rather than a doctrinal one based on the Spirit's own inspiration. "Canon" in traditional thought refers also to the use of Scripture to judge the thoughts of man. "When ancient Christian texts speak about 'canonical' books, it is often their normative quality, or the authoritativeness, of certain books that is intended, rather than their inclusion in a particular list of titles."[1] Scripture is not a list assembled by man, but the ruling authority for the Church and the external touchstone of truth, which judges all human words.

It has become fashionable to say that the *antilegomena* (the biblical writings partially spoken against in the Early Church) are of "secondary authority."[2] What does a divine "secondary authority" actually mean?

---

[1] Einar Thomassen, *Canon and Canonicity: The Formation and Use of Scripture* (Copenhagen: Museum Tusculanum Press, 2010), 9.

[2] This paper cited is controversial, but it is used here as illustrative of broader themes within modernism and the LCMS, not to single out the author. Its authorship has been affirmed and none of its theological positions to date have been retracted. The dissemination of false doctrine requires refutation. The milder published version of this essay is cited where it follows the original. Jeffery Kloha, "Text and Authority: Theological and Hermeneutical Reflections on a Plastic Text," Listening to God's Word, Nov. 7–9, 2013 in Oberursel (unpublished paper; http://steadfastlutherans.org/wp-content/uploads/2014/02/Text-and-Authority.pdf), 12; J. Kloha, "Theological and Hermeneutical Reflections," in *Listening to the Word of*

Can God contradict Himself enough for sinful man to distinguish between primary and secondary *divine* authorities? God Himself is not divided so easily. The formation of the canon is hugely problematic for modern theologians who deny inspiration and all divine authority. "A single, normative text is entirely unhistorical," only because of a critical definition of history.[3]

> My concern is that to speak of a single act of inspiration (in much the same way that Islam describes the Koran) that produced a set of single normative books that now comprise the 'Bible' without fully acknowledging and dealing with the historical evidence leaves us vulnerable to rhetoric which denies the authority of the biblical text. But God works in history.[4]

No hard evidence is offered; instead, the uniqueness of Scripture, because other religions claim to have inspired writings, is denied by default. So God's book must be like any other writing authored by sinners, because that is reasonable to evil-doers.

The problem of authority is acute where the direct, divine authority of Scripture's words is disallowed. Classically, the canon was not a critical problem, but a supernatural gift recognized only with the same Spirit who spoke through the prophets and the apostles. God's undiluted Word alone, not human methods or investigations, establishes beliefs. Any theological conclusions based on distinctions between levels of God's words are *de facto* a denial of *sola Scriptura*. A word is either spoken by God or not: there are no half-divine, half-inspired words with one half the authority. The Word of God is spoken by God and remains forever, despite how poorly it has been received.

"The canon, of the *homologoumena* [the writings mostly accepted] and *antilegomena,* is a historical problem, not an article of faith."[5] An inspired, exhaustive list of inspired books is not found in Scripture itself, so there can be no divine authority for any such list. Only on the basis of external evidence, such as the human author and date of writing, can the books be analyzed. However, no division can be made between

---

God: Exegetical Approaches, ed. A. Behrens, 196.

[3] J. Kloha, "Text and Authority," 16.

[4] J. Kloha, "Theological and Hermeneutical Reflections," in *Listening to the Word of God: Exegetical Approaches,* ed. A. Behrens, 196.

[5] R. D. Preus, *Inspiration of Scripture,* xi.

inspired books, which only the Spirit authorizes. No critical judgment based on any amount of evidence can make human words God's words. In the almighty historian's catbird seat, the Scriptures must have been edited and altered over time, like the "drafts of the Gettysburg address," but Scripture furnishes no proof, and external criteria judged by man do not establish the divinity of God's words.[6] Any pre-inspiration activities, such as St. Luke's historical investigations, cannot be played against divine authorship—unless God Himself is excluded totally from His world. It is impious and blasphemous to judge divine revelation on the basis of human witnesses, if God truly exists and is of a higher authority than man. A god that is authorized by man is not the Lord Christ.

Though it is often implied, the *antilegomena* biblical books are not of less divine authority than the wholly accepted *homologoumena*. That distinction tells us about the people reading and using Scripture, not Scripture itself. Either a book is God's Word or not. There is no half-human, half-divine monstrosity.

Why dispute the human witnesses and be concerned with human authorship? The Spirit used human authors, including their experiences, intelligence, and vocabularies. Because the Spirit gave the words in a real, historical event, those external attributes can be discussed and used positively. While no historical investigations can establish the Bible's divinity, or deny it, apologetically, there is much value in the earliest human witnesses. Those with God's Word are not afraid that Scripture will contradict any true fact. To impute to God's Word uncertainty or a secondary status is blasphemy.

We have discretion in what texts we use in our theological arguments. An article of doctrine can usually be defended from multiple texts. One word of God is enough for faith. Every relevant text is not needed to establish a divine teaching. This choice to refer to the most generally accepted books (such as the *homologoumena*) is based on limits of weak men, not God's Word. This was done to ensure that the divine teaching is beyond reproach for those who might (erroneously) doubt certain books. To voluntarily restrict oneself in a controversy is an act of love—not a condemnation of God's words. After all, it is the Spirit who authorizes His supreme Word, not a list made by a sinner.

---

[6] J. Kloha, "Text and Authority," 14.

One may freely refrain from using parts of Scripture that others do not consider authorized. This is not a judgment over God's Word, but the acknowledgment that even a limited portion of God's Word can settle doctrinal disputes. Salvation is not to be put in jeopardy for any man. God's Word is more than a list—it brings Christ to save.

No man can force another to accept something as God's Word—that is the Spirit's work. Faith is a divine conviction, not something based on reasonable proofs. The goal of the *homologoumena* distinction was to make doctrine *more* sure, not to critique God's speech by human criteria. This distinction describes the voluntary restriction of scriptural proof for contentious issues. If man authorizes the scriptural books, they have only human authority and need not be accepted as Christ's Word.

Did its later historical acceptance change whether Scripture was given originally by the Holy Spirit? No, it is either from God or solely from man. Contrary to moderns who make much of the Early Church's reception of the New Testament books, it is not a significant matter at all. It can even be safely ignored because Scripture did not ultimately come from any man: "For no prophecy was ever produced by the will of man, but men spoke from God as they were carried along by the Holy Spirit" (2 Pet. 1:21).

Jesus and the earliest Christians were content to promulgate the faith based only on the Old Testament. Man's reception or rejection of a revelation cannot establish it or invalidate it. Inspiration—the Holy Spirit Himself—defines the canon, if the canon is defined as God's Word. The main question that must be answered today is not the scope of the canon, but whether it is of human or divine authority. No human reception or evidence makes man's words divine. This statement of a pope who died in A. D. 604 says that no church has a role in authorizing Scripture:

> These [books of Scripture] the Church holds to be sacred and canonical, not because, having been composed by simple human industry, they were later approved by her own authority, nor merely because they contain revelation without error, but because, having been written by the inspiration of the Holy Spirit, they have God for their author and were delivered as such to the Church.[7]

---

[7]Gregory the Great, quoted in: S. W. Hahn, "For the Sake of our Salvation," 31.

Controversies must be settled by the Word of God, without any occasion to doubt. Luther's comments are well known: the book of James is "not the writing of any apostle."[8] They sound modern, but only because perhaps we too readily accept the canon based on human authority. He critiques not the Word of God, but the human word of a human author who was *not* an apostle. What is not apostolic is not Scripture, nor inspired, in his mind. Notice that he did not foist his view on later Lutherans. Very few have followed him in this matter. It did not alter his doctrinal conclusions one iota, showing this to be a minor issue in actuality. In fact, Cardinal Cajetan, the opponent of Luther, also doubted the book of James.[9] What Luther claimed may readily be disputed, but he said nothing about Scripture proper. If he spoke about inspiration, it was only the inspiration of a pseudo-apostle, not the Spirit's chosen instrument.

What is an alternative to inspiration? "The recognition of 'Christ crucified' as *the* criterion of canonicity helps explain the standard by which the early church identified which writings were apostolic."[10] The aspiration to judge God's words by any criteria would have been anathema to pre-moderns—patent unbelief. Their main concern was to determine which apostolic writings were inspired of God, and therefore God's literal words. Apostolic words are simply the words of men, unless God the Spirit spoke by them. "Christ crucified" is just two words. Two words are not an authority, but a (rather limited) description of the Gospel. Is the Arian Christ, the Ebionite, or the Mormon Christ authoritative? In modern theology Christ becomes just a non-authoritative idea, a critical principle to rummage through every verse of Scripture.

One starts either with the assumption of the divine unity of the given *words* or an "awareness of the disunity of the biblical canon and the multiplicity of biblical theologies."[11] Disunity, based on the premise of multiple authors, is supposedly more scientific when studying what claims to be a revelation from God. "Modern exegetical research

---

[8] J. M. Reu, *Luther and the Scriptures*, 24.

[9] J. A. O. Preus, "The New Testament Canon in the Lutheran Dogmaticians," *Springfielder* 25:1 (1961), 134.

[10] P. H. Nafzger, *These Are Written*, 124.

[11] Roy A. Harrisville and Walter Sundberg, *The Bible in Modern Culture: Baruch Spinoza to Brevard Childs* (Grand Rapids: Eerdmans, 2002), 11.

speaks for the plurality of theologies in the New Testament."[12] It only allows human words and authors, so the disunity of Scripture is a given. Modern interpretation speaks volumes about method-users, who each have their own individual authorities—themselves!

Luther, however, was free to judge Christologically what he held to be non-divine, precisely because he had reliable knowledge of the Christ he knew *from* Scripture. The modern position puts man above God in allowing him to determine what is worthy of God: "The authority of the documents in the canon comes from their apostolic content, not from their being in the canon. The canon simply acknowledges their apostolic content."[13] Content is critically judged, but inspired words must simply be accepted as authoritative. Luther's knowledge was based on a divine conviction, which is not a critical principle, but a certainty that God has spoken. "All heresies and errors in connection with the Scriptures have arisen, not from the simplicity of the words, as is almost universally stated, but from neglect of the simplicity of the words, and from tropes or inferences hatched out of men's own heads."[14]

Lutherans are defined by not defining the biblical books in a list. While Rome, the Eastern Orthodox, and most Protestants have a definitive list, Lutherans saw establishing divine authority by man's authority as completely backwards. Scripture does not need man to speak for it—it is clear enough. Only God Himself can convince man. The Holy Spirit in His clear Word is His own proof, not sinners. Calvin, like Luther, "dismissed, as making the Holy Spirit a laughingstock [*ludibrio*], the supposition that the church must decide for us which books should be received as canonical."[15] These pre-moderns clearly saw that human authority was no substitute for God's own authority.

It is not that parts of Scripture are less certain, but that weak men are less certain about the divine status of particular books. The real contention with the *antilegomena* is the lack of historical witness identifying the human author. These are all issues relating to the secondary author, not the primary. No historical issues are problematic

---

[12] M. Ruokanen, *Doctrina divinitus inspirata*, 145.

[13] Charles A. Gieschen, "The Relevance of the *Homologoumena* and *Antilegomena* Distinction for the New Testament Canon Today: Revelation as a Test Case," *CTQ* 79:3–4 (2015), 285.

[14] *The Bondage of the Will* (1525), LW 33:163.

[15] Quoted in: *Inerrancy and the Church*, ed. J. D. Hannah, 170.

for the eternal Spirit. Canonicity is not proved by external criteria, but only by the internal testimony of the Spirit.

> Faith, which considers the testimony of the primitive church which witnesses that these books have been written by apostles and evangelists, is a human and historic faith; but faith, which believes that this or that book is divine and canonical, or comes from the Holy Ghost, is divine faith, and this does not rest on the testimony of the church, but on the internal criteria of Holy Scripture and primarily on the testimony of the Holy Spirit.[16]

To finalize or authoritatively define the scriptural canon is not within man's power. The book of Revelation, however, is a strong testimony to not expect more words.

Modern man sees legitimacy only in rational proof. He is convinced by arguments which appeal to reason and its certainty. "The criterion of inspiration falls before the stony unbelief of modern criticism and the demand for scientific proof."[17] This modern approach to knowledge is part of our makeup. But human authority allows no real certainty. The reality of the Spirit as a divine deposit in the believer explains the fact of the Reformation. Boldness in confessing Christ is from the Holy Spirit: "they were all filled with the Holy Spirit and continued to speak the word of God with boldness" (Acts 4:31). Even the most fundamental Christian confession requires not sound reason but the person of the Holy Spirit: "Therefore I want you to understand that no one speaking in the Spirit of God ever says 'Jesus is accursed!' and no one can say 'Jesus is Lord' except in the Holy Spirit" (1 Cor. 12:3). The intellectual power of the scholar does not replace the power of God's Spirit in the believer.

While it is claimed that "inspiration short-circuits the historical process in the formation of the canon," it does no such thing.[18] Inspiration says there is more than inaccessible history that man sits over in judgment. God the Spirit's involvement is not a negative, but the only thing that allows there to be an actual Word of God. Only the bold rationalist sees God's interaction with man as a problem. External

---

[16] Johannes Andreas Quenstedt (d. 1688), quoted in: J. A. O. Preus, "The New Testament Canon in the Lutheran Dogmaticians," 148.

[17] J. A. O. Preus, "The New Testament Canon in the Lutheran Dogmaticians," 149.

[18] P. H. Nafzger, *These Are Written*, 133.

criteria surrounding the human authorship have always been held as complementary to Scripture's divine inspiration. What kind of god is it that must be kept far away from man?

What the denial of inspiration does is to short-circuit any way to obtain divine truth and judge human reformulations of it. "The Christian Church is indeed older than Holy Scripture, that is, older than the written Word of God," but not older than God's words, which are eternal.[19] The canon is not decided by human criteria or reason. "The whole canon is the context for each part of it," because it ultimately has the same Author.[20] Scripture's unity is organic and an article of faith—not something foisted on it from outside by man. That is why it does not matter if certain books are unattributed to a human author, such as the book of Hebrews. They are received by the Christian on the basis of their own divine witness, not the fallible, errant authority of man. One of the main contributors to the Formula of Concord summarizes the true Lutheran position: "The canonical Scripture has its eminent authority chiefly from this, that it is divinely inspired, 2 Tim. 3:16, that is, that it was not brought forth by the will of men but that men of God, moved by the Holy Spirit, both spoke and wrote, 2 Pet. 1:21."[21]

---

[19] F. Pieper, *Christian Dogmatics*, 1:193.
[20] Martin Naumann, "Messianic Mountaintops," *Springfielder* 39:2 (1975), 8.
[21] Martin Chemnitz, *Examination of the Council of Trent*, 4 vols., trans. Fred Kramer (St. Louis: Concordia Publishing House, 1978), 1:176.

Chapter 23

Divine Doctrine is the Correct Presupposition

Everyone starts with a presupposition vis-à-vis God's Word, as self-labeled "post-modernists" would point out. However, it is not human understanding or cultural and philosophical values which matter before the holy God. The divine doctrine, revealed in Scripture itself, gives the right approach, but this is only learned through the Spirit's illumination. The sinner receives Christ not through a mystical experience, but in the Gospel's forgiveness, communicated in human words.

Dogmatic approaches are considered unhistorical because the Lord of time has been walled off in eternity. But it is not man who makes and sets limits for God. In fact, Christ is the basis for history, as the Creator of the world and time itself. His communication is the only authoritative, factual history.

Even in science, a faulty hypothesis will lead to incorrect conclusions. Data is mishandled by a bad method. Not every theory or assumption is correct. That is the great beauty of real science. It allows for judgment to be passed based on tangible evidence that may be touched and handled by all, unlike philosophy. The results of historical criticism speak for themselves: " 'After the Socinian criticism [c. 1700], no honest and clear thinker' attempted to renew these dogmas 'in a literal sense'."[1] Modernism is the conceit that man is above any supposed divine teach-

---

[1] Wilhelm Dilthey, quoted in: K. Scholder, *Birth of Modern Critical Theology*, 44.

ing, as a result of all his progress. Yet empowering critical thoughts do not save from death and hell. At Judgment Day no one will be able to ignore Christ's plain words.

In theology God alone, not the earthly result, speaks with authority. He says that faith in His Word, while appearing to be divine slavery, is truly blessed freedom. "For freedom Christ has set us free; stand firm therefore, and do not submit again to a yoke of slavery" (Gal. 5:1). To submit to the authority of Christ is freedom from death and sin. The world's idea of freedom is actually slavery to sin, because transgressors can only rebel against God's law. Being able to think all thoughts and act against divine mandate leads only to death.

The truth of the Gospel cannot be verified by man. In its naturalistic religiosity, the main presupposition of modern exegesis is that human understanding is the highest authority, and therefore practically divine. It rules out hell, the main negative outcome of denying Christ's Word.

What God speaks is true, though it cannot be verified now. Every unbeliever will eventually observe the truth: "at the name of Jesus every knee should bow, in heaven and on earth and under the earth" (Phil. 2:10). They will not bow in the freedom of the Spirit—it will be too late for salvation.

What controls the method of interpreting Scripture? Either sin-bound man rules, or Christ through His words. An absolute method supposedly allows man freedom from divine authority. The Word of God does the opposite. It exercises its authority by bringing man to Christ through the divine teaching, making him new and spiritual. The Gospel frees on account of God's action in heaven, not man's mental powers. The rigid, unbending teaching of God is actually complete freedom from death and Satan's rule. Since it is not man's creation, it can be trusted. It brings out of darkness into the marvelous light of Christ. Man is freed from trying to be an objective, all-knowing deity and can simply be a listening child of God. Authority is then something good, because of the one it emanates from—the crucified Lord.

The first profit of Scripture listed in 2 Tim. 3:16 is doctrine. The only way for theology not to be anthropology, merely statements by men about men, is through the principle of Scripture. The only "proper principle of theology is divine revelation," which we find today only in the Bible.[2] So what is not biblical is not theological, since it is not revealed by Christ. A human understanding of history or science is not the main purpose of Scripture, but "when Scripture incidentally treats scientific subjects, it is always right."[3] Even the most mundane earthly matters in Scripture are rightly held as divine, so that the precious doctrinal gems of Christ are untarnished by man's reason. Christ's authority has no limit. An errant Scripture means errant doctrine, which does not permit an inerrant Christ. Who would trust in a Christ who cannot make Himself known clearly or even makes mistakes? A Christ who does not have authority over secular matters is not truly God.

It is man that is sinful and unable to think rightly about God or His works. "Our teaching, however, straightaway denies that any mortal is by nature suitable for perceiving [the mysteries of Scripture], because all men are creatures that do not perceive the divine."[4] Man cannot come unprejudiced to God's Word. He should be taught the divine approach in faith according to what God has revealed. This is now called "indoctrination," a negative term for moderns.

Doctrine is not just an abstraction useless for real life; it is the way in which God reveals Himself. Sleeping on the Bible or carrying it around does not bring faith, but what Scripture expresses in human words is fully divine. God reveals Himself in doctrine. To have and believe the right doctrine is life itself. Christ has only promised to come through scriptural teachings. When the literal meaning of Scripture is lost, "really good theologians are no longer produced. Only the true principal meaning which is provided by the letters can produce good theologians."[5]

There is no complete knowledge or scientific objectivity, for we know dimly (1 Cor. 13:12). But what has been revealed is for our good. Scripture serves a negative function, which the Gospel cannot, by ruling out erroneous teachings. Doctrines are propositions, after all, extracted from Scripture.[6] If no propositions bind theologians, they can proclaim

---

[2] Johann Gerhard, quoted in: F. Pieper, *Christian Dogmatics*, 1:59.
[3] F. Pieper, *Christian Dogmatics*, 1:75, 317.
[4] M. Flacius, *How to Understand the Sacred Scriptures*, 89.
[5] Luther, *Answer to Goat Emser* (1521), LW 39:178.
[6] D. M. Farkasfalvy, *Inspiration and Interpretation*, 161.

whatever they see fit. Freedom of this world leads to anarchy. The move away from the literal authority of Scripture is evil. Like all other rebellions, it is "the Satanic revolt against divine authority."[7]

"The force of [Jesus'] teaching often depends on [the Scriptures'] literal truth."[8] Making authoritative statements is vastly different from being systematic and filling every rational gap in a perfectly coherent system. "God is no systematician and the Bible he gave us is the most unsystematic book ever written," to the unbeliever.[9] Despite the observations of rebellious reason, God is systematic in the highest way as the author of all truth. But this truth is to be believed, not kept at a critical distance. The term "dogmatics" fits the Scriptures more than "systematics." "To erect a system is incompatible with theology based solely on the Scriptures, which is not a system in the strictest sense."[10] There are definite gaps and limits to what can be asked of it. The problem is never with Scripture, but that sinners do not fully understand or accept the doctrine taught in it.

Saying one believes in a norm is not the same as actually norming one's personal speculation. Sinners can only err when thinking about God, unless they follow His revealed Word. "The articles of faith are not only above, but contrary to, corrupt reason, which judges them to be foolishness."[11] Only by God's direct activity, which will involve suffering of some sort, can man speak on the basis of God's Word. Man is a slave to sin and the Devil until He is freed. He cannot understand the divine contents of Scripture, but only the outward meaning of the words. An unbelieving critic who denies the clarity of Scripture is the same as a completely blind person denying that the sun shines. "It is most absurd to deny the perspicuity of some saying or doctrine just because this or that shoemaker, gardener, tanner, or someone else steeped in five contradictory heresies has said that he does not understand it clearly."[12]

To know and believe God's Word is to know God Himself. The very

---

[7] Michael Bakunin (d. 1876), quoted in: W. A. Visser 't Hooft, *The Fatherhood of God in an Age of Emancipation* (Philadelphia Press: Westminster Press, 1982), 99.

[8] *Inerrancy,* ed. N. L. Geisler, 2.

[9] Herman Sasse, Letter to Dr. J. A. O. Preus, Feb. 24, 1970 (Copy in Concordia Theological Seminary Library, Fort Wayne, IN).

[10] A. Hoenecke, *Evangelical Lutheran Dogmatics,* 1:xiii.

[11] F. Pieper, *Christian Dogmatics,* 1:199.

[12] A. Hoenecke, *Evangelical Lutheran Dogmatics,* 1:478.

words of God come to us in only one form: Scripture. God is not limited to writing, though He has limited us, out of pure grace and divine mercy, to a single source of truth, so that we may be sure. Jesus is the truth, for He said: "I am the way, and the truth, and the life. No one comes to the Father except through me" (Jn. 14:6). However, "we only know Him in the instrumentality of *truths*."[13] The scriptural Jesus spoke objective truths with His mouth. The Scriptures are Christ's mouth today.

The Spirit states in Scripture: "Therefore whoever disregards this, disregards not man but God, who gives his Holy Spirit to you" (1 Th. 4:8). Whoever disregards this divine information about abstaining from sexual immorality disregards God. It proves that having Christ is contingent on having a teaching, or at the very least not denying it. "This" in 1 Thessalonians 4:8 refers to something that translates to every time and culture, that is, a true and universally valid proposition. There is no alternative: "The sole beginning of theology is the word of God; therefore what is not revealed in the word of God, is not theological."[14] These divine truths are not optional. "Who does not know God, just as He revealed Himself in His word, he strays from the true God; who ignores or denies the definition of God handed down in Scripture, he ignores or denies God Himself."[15]

Since God is only known in His Word, a non-doctrinal reading or hearing is unbelief, and therefore sinful. It is revealed doctrines which create faith and bring Christ. "The Scriptures must be read in the fear of God," as if sitting before the Lord of heaven and earth.[16] While "the old morality [orthodoxy] celebrated faith and belief as virtues and regarded doubt as sin," in the Cartesian morality doubt is the highest virtue.[17] The question is not whether it is old or recorded by men, but *what* God actually says and teaches in His Word. Since Christ is fully God and one with the Spirit, "to disbelieve or disobey any word of Scripture is to disbelieve or disobey God," that is, Christ Himself.[18]

---

[13] B. Ramm, *Special Revelation and the Word of God*, 117.

[14] Johann Gerhard, quoted in: Johann Wilhelm Baier, *A Compendium of Positive Theology,* ed. C. F. W. Walther, trans. Ted Mayes (unpublished draft, 2012), I.25; 1:64.

[15] Johann Gerhard, quoted in: Baier, *Compendium of Positive Theology*, I.32.e; 1:81.

[16] M. Flacius, *How to Understand the Sacred Scriptures*, 69.

[17] V. A. Harvey, *The Historian and the Believer*, 103.

[18] Wayne Grudem, quoted in: N. C. Grubbs and C. S. Drumm, "What Does

Theology Have to Do with the Bible?," 68.

Chapter 24

Proof-Texting: The Only Theological Sin

Doctrine is incomprehensible to modern man as an external authority. Scriptural teachings are only historical ideas and concepts open to investigation, not divinely binding decisions that demand that man hear in a submissive position. Without the inspiration of the Spirit undergirding the Scripture's words, timeless doctrine is an impossibility. The authorization for divine teaching can only come from divine words. Even in the years following Luther, "Protestantism, then, did not bring any immediate relief from the old tendency, as in the days of Tertullian and St. Aquinas to measure up all science and philosophy against the relevant statements of Scripture and Genesis in particular."[1] The nature of Scripture, based on its own specific testimony, not man's neutral observations, demanded it. Luther comments: "If we could believe that God Himself speaks to us in Scripture we would read it with all diligence."[2]

This critic is right: the "primitive dogmatical way of arguing . . . tacitly includes a fundamental rejection of the critical historical method."[3] Moderns derisively call appealing to divine words "proof-texting." Cit-

---

[1] J. Y. Simpson, *Landmarks in the Struggle between Science and Religion*, 139.
[2] Quoted in: Siegbert W. Becker, "The Word of God in the Theology of Martin Luther," unpublished paper (1963; http://www.wlsessays.net/bitstream/handle/123456789/363/BeckerTheology.pdf), 11.
[3] G. Ebeling, *Word and Faith*, 21.

ing a divine citation that destroys all scientific objectivity is the only theological sin. Yet prior to the Enlightenment, "divine revelation was consulted as the final arbiter of truth."[4] Because of the denial of the origin of God's Word, " 'How do you know' is the first question of theology."[5] That is a great question, but it cannot be answered correctly without a direct Word of God, above all man's words, that speaks to the topic.

Authority is found in words and tangible communication. There are no ideas or truth without words. "Any simple, direct use of scriptural statements in formation of theology is impossible for men today."[6] If theology is to be more than a man-made enterprise, it must speak with divine authority. The only way for that to happen is to use Scripture to criticize every man-made thought and to take captive human speculation, which can never arrive at the truth of the Gospel. Without possessing the words and thoughts of God, there is no true theology. Divine "theology is simply the arrangement of Biblical facts," or it is man-centered speculation.[7]

Historical criticism liberated "believers from having to accept every passage of Scripture literally or the entire text as the Word of God."[8] It did not discriminate, but obliterated all literal authorities and religions. The Bible, for the modern, is still valued as a resource, but not an absolute authority to be obeyed. Scripture has become a Rubik's Cube to scholars who simply want the freedom to play with it as a neutral object. This is only possible by "denying the relevancy of standard proof-texts," "supposing that religion could exist without a doctrinal foundation."[9] These were radical ideas when they were disseminated by Johann Salomo Semler (d. 1791), but they now constitute the greatest attack on God's Word. The authority of the Bible is spoken of with reverent tones, but the authority relates to the whole as construed by the critic, not particular words. Modern man is *the* authority, even in

---

[4]S. J. Grenz and R. E. Olson, *20th-Century Theology*, 17.
[5]C. E. Braaten, *History and Hermeneutics*, 12.
[6]C. H. Pinnock, *A Defense of Biblical Infallibility*, 5.
[7]A. H. Strong, *Systematic Theology*, 21.
[8]J. C. Livingston, *Modern Christian Thought*, 240.
[9]*Library of Universal Knowledge: A Reprint of the Last Edinburgh and London Edition of Chambers's Encyclopædia*, vol. 6 of 15 (New York: American Book Exchange, 1880), 621.

reading God's Word.

This method of interpreting Scripture as a whole is paramount for preserving unbelievers from the full authority of God. The interpreter grants authority to the individual words, not the other way around. In the pre-modern view, a false teaching is an attack on the words of Scripture—God's verbal commands. To deny a single doctrine is to deny God, so no passage, understood correctly, could be denied. Since God is the primary author, all contents of Scripture, in particular and collectively, are the revealed Word of God. Luther saw that Scripture was being denied in his day, but he took heart by singing "The Word They Still Shall Let Remain." God's words, not just His Word, remain eternally valid. "The formula 'It is written' means that this passage in the Old Testament is the [writing] of God and therefore is the truth of God" for all time.[10] What the writing says, God says.

While the temptation is great to defend the possibility of revelation, that remains in the realm of reason-based philosophy. It is not a Christian activity, but remains speculative or apologetic. It is not reasonable that God wrote Scripture, or rather, Scripture will not be reasonable to sinful man. Yet the fact of revelation remains. Scripture should be taken on its own terms. It cannot be studied from above as a human object without sinning against the Lord. It is to be received as an oracle in which God speaks, not examined scientifically like a microorganism.

One of the early popularizers of historical criticism stated that "the fundamental cause of [Christian] mischief lies in the fact that the Bible is not read like any other book."[11] Since only God's speech authorizes absolute truth, the literal meaning of the Bible is not found by a reductionist method or man's private judgment. Rather, the literal meaning "is that which the Holy Ghost doth in that place principally intend."[12] The meaning of a passage can be literally figurative—if the Spirit spoke in a figurative sense—as the Spirit's plain, intended, stable meaning. The doctrine of inspiration does not prescribe a rational method for interpretation.

---

[10] B. Ramm, *Special Revelation and the Word of God*, 165.

[11] Benjamin Jowett (d. 1893), quoted in: J. C. Livingston, *Modern Christian Thought*, 238.

[12] John Donne (d. 1631), quoted in: P. Harrison, *The Bible, Protestantism, and the Rise of Natural Science*, 109.

Divine, saving knowledge must be revealed and communicated. The Spirit does the work of enlightening, so that the Word is understood. Through the Word He brings Christ's righteousness and faith. These are not separate activities, since faith comes by the Word: "If you abide in my word, you are truly my disciples, and you will know the truth, and the truth will set you free" (Jn. 8:31–32). "The goal of exegesis is to lead to confidence, assurance, and an increase in faith," not agnostic knowledge.[13]

God does not lead astray in any area of revealed knowledge. An errant god is certainly not the Christ who has all authority in heaven and on earth. He is truthful in whatever He speaks. The scriptural Christ knows the creation He spoke into being over six days better than the scientist.

Why are proof-texts important? They are divine proof and authority above man's flimsy thoughts. "There is no doctrine of Christian theology which is not set forth in unmistakable terms in some text of Scripture."[14] This is not a limitation of Scripture, but a recognition that Christianity is of divine origin. The words are divine, so that "every passage of Scripture was potentially a microcosm in which the meaning of the whole [of Scripture] could be enfolded."[15] There should be a limit on what man, not God, says. Proof-texts are necessary for establishing the doctrinal foundation, but hardly exhaust the use of the Bible. Orthodox sermons and devotions attest to that fact.

Without a specific divine warrant, nothing is certain. In the man-centered discovery of truth, "no historical judgment, however certain it may appear, ever attains any more than probability."[16] Faith is based on God's words, specifically the assurance found in divine promises. Lutheranism is built on proof-texts. The Small Catechism grounds the human words of Luther in specific divine passages. It asks for proof in a citation: "which are these words and promises of God?" Only if the words are divine is the discussion closed and certainty assured: "This is most certainly true." A critic, who has the modernist spirit, cannot

---

[13] T. Engelder, *Haec Dixit Dominus*, 5.

[14] T. Engelder, *Haec Dixit Dominus*, 22.

[15] P. Harrison, *The Bible, Protestantism, and the Rise of Natural Science*, 47.

[16] Wilhelm Hermann (d. 1922), quoted in: J. C. Livingston, *Modern Christian Thought*, 284.

utter this confession.

Criticism has given "shock to the old unquestioning faith, which holds [Scripture] to be above all human judgment, and is content to listen to every sentence as a complete Word of God."[17] Neo-orthodoxy (Barthianism) despises literal proof-texting as anti-intellectual. This leads to a wordless Christianity. It is "neither new nor orthodox."[18]

Reading God's Word should not be a critical decision based on the neutral observation of all the relevant evidence. No one decides to believe or understand spiritual things. The "Holy Ghost is given, who works faith; where and when it pleases God, in them that hear the Gospel."[19] This Spirit-wrought faith does not mean the believer will understand all of the Bible according to the modern notion of scientific completeness or perfect coherence. Rather, it is reliance on a Word of God, a divine promise which forgives sins. This God who justifies sinners does not mislead in any matter—to claim such is an attitude of unbelief. Faith can say, without empirical proof, that Scripture is "the only book in which historical errors cannot occur."[20] Following the tack of Justin Martyr, inerrancy is a humble confession of faith before the Maker of heaven and earth: "If any apparently contradictory passages be adduced, he will rather confess that he does not understand them."[21] Either man proves by rational means or God proves by His own authority.

For the despisers of the bare, individual words of the Spirit, what is the alternative? "Christ, the person," which sounds very spiritual. But "Christ," separated from Scripture, as the sole judge and norm of the content of theology, is actually philosophy. The Gospel is not a rational principle, but a true history—a story that contains valid, transferable, and timeless propositions. "The neoorthodox located the objectivity of the Word of God uniquely in the Person of Christ."[22] Yet as a historical event, criticism allows no access, especially not in a Word which demands uncritical obedience. But if Scripture is only an empty vehicle for the Spirit, much like the Reformed view of the Lord's Supper, man gets to fill the gap of knowledge left unexpressed. Christ

---

[17] H. D. McDonald, *Theories of Revelation*, 2:230.

[18] A. Hoenecke, *Evangelical Lutheran Dogmatics*, 1:169.

[19] AC V:2–3.

[20] F. Pieper, *Christian Dogmatics*, 1:318.

[21] A. A. Zaun, "A Study of the Idea of the Verbal Inspiration," 78.

[22] R. C. Sproul in *Inerrancy*, ed. N. L. Geisler, 349.

becomes a wordless ideal about which modern man speaks many words.

Proof-texting is not just "proving." It acknowledges that Scripture, therefore God Himself, is the only source of divine teaching. As a result of sin, whatever "doctrine does not originate in the Scriptures . . . can be the doctrine of no one but the devil."[23] Not even reasonable conclusions can hold weight, if they go beyond the authority of the divine words. Luther made a statement in 1517 that already showed the error of the Reformed method to come: "No syllogistic form holds in divine terms."[24] Sinful reason cannot even make sound deductions from revealed principles. All certain knowledge must be directly revealed, and believed unconditionally.

Without submission to the text of God's true revelation, orthodoxy is impossible. The label of "fundamentalism" can only be the sign of persecution to those claiming to possess truth. However, confessing that Scripture is God's Word is not the same as possessing a full understanding of it or the ability to rationally defend it. It is believed because Christ comes and saves in His words.

To speak with authority over educated men is forbidden by moderns. The rationale is that cultured man has achieved powerful evidence of his critical authority. He has technological mastery and is scientific lord over the world. Peace is the direct result, it is assumed. But the allure of the modernist religion does not depend on completeness or facts. It claims to define truth as a moldable thing.

Scripture as the sole *principium* [principle and fountain] for truth does not compute for a modern. How can a book be a principle? In modern thinking a principle implies a rational process by which man deduces truth. But *sola Scriptura* is not a starting point whereby man only gets an initial direction. It is the confession that all sure knowledge must be from God—everything else is to be doubted. Scripture is not a theory to be tested, but the actual rule and judge of all theology. Since it is God's book, established by His own witness, all contradictions with any other true fact are ruled out *a priori*.

Scripture, as a book, is not to be adored or praised, but used and obeyed. Its words evaluate the truth of claims of theology proper.[25]

---

[23] Luther, *Avoiding the Doctrines of Men (1522)*, LW 35:137.

[24] Quoted in: K. Scholder, *Birth of Modern Critical Theology*, 168.

[25] *The Bible in the History of the Lutheran Church,* ed. J. Maxfield, 20.

The Bible is not just a resource or guide to salvation, it measures all statements concerning God. The truth is not ineffable. It can be stated and communicated, so the Spirit may awaken men from unbelief. This is not an easy task; in fact, our own Scripture says it is closed to sinners. "With man this is impossible, but with God all things are possible" (Mt. 19:26).

But what to make of passages that are personally difficult to understand? Augustine put the best construction on the divine Word:

> The Holy Ghost has arranged Holy Scripture in such a magnificent way that through the clear passages He appeases hunger and through the dark passages He prevents loathing. For hardly anything is derived from those obscure passages but what is stated elsewhere most clearly.[26]

It is precisely because man cannot objectively examine Scripture, and is not lord over it, that it is so dear to Christians. God Himself speaks to the Christian in its words, and they continue to teach man spiritual truths. No matter how much it is used it does not become overused or trite.

The true interpreter of Scripture is the Spirit working though His own words. "In the exposition of Scripture and in the process of true learning, after the gift of the Spirit of God, the comparison of similar portions of Scripture, whether words, phrases, or subject matter, is the most effective tool."[27] The modern wants an innovative use of Scripture—a principle that allows freedom for reason. But the Christian should go to the places where a topic is addressed by the Spirit, not those it is not. Locating the most crucial passages is often the hardest part of interpreting, not the words themselves. Scripture is quite simple. Understanding the Spirit's wisdom does not require a great intellect or worldly learning.

It is not just one perspective that Scripture offers, but the whole and undefiled truth of God. This is the only way to relate the scriptural documents of vastly different ages and human authors. "The ultimate literary context of any given passage is the whole canon of Scripture."[28] This does not mean a literalistic understanding, as is often put forth. "This term *sensus literalis* [literal sense] refers to the meaning intended

---

[26] Quoted in: F. Pieper, *Christian Dogmatics*, 1:324.
[27] M. Flacius, *How to Understand the Sacred Scriptures*, 87.
[28] G. Goldsworthy, *Gospel-Centered Hermeneutics*, 200.

by the author, whether figurative or non-figurative."[29] God's voice is the final authority for the Christian. Whatever Christ does not speak is not Christian. "Here once again [in 1 Tim. 4:1–7] you see that God himself through the mouth of St. Paul ascribes to the devil doctrines of men. . . . Whoever is not satisfied with this text, what indeed will satisfy him?"[30]

The strongest evidence of proof-texting is the Lutheran position on the Lord's Supper. While moderns find uncertain and non-propositional themes alluding to the Sacrament in every biblical nook and cranny, Luther bases his doctrine on one word: "is." He knew that the passages which divinely institute the Supper are the only ones which could give certain knowledge. For Him, it was no different than hearing God speak from heaven. This is why his position did not change based on the Roman or Reformed opinions: he had the verbalized truth of God. Basing a doctrine on a single, inspired word is infuriating for moderns and ecumenical pacifists, but Luther was neither. "It is unfortunate, however, that Luther did not seek to interpret this passage [of the Supper's institution] a little more freely in the light of his elsewhere expressed, and most cherished, fundamental principles."[31] His principle was Scripture and he did not dare to deviate even a jot or tittle from the divine Word. At the Marburg Colloquy Luther asserted most anti-rationally: "If God commanded me [via the Scriptures!] to eat dung, I would do it."[32] Luther did not try to find vague sacramental eating motifs in Scripture to prove his position. He had the most certain nude words of God—which are just as clear today.

Proof-texting does not mean treating every passage the same. Scripture itself (meaning God) must determine the place and relevance of each part. "The Evangelical Lutheran Church assigns to every doctrine of Scripture the rank and significance which it is given in God's Word itself."[33] Outside sources may humanly support the truth established

---

[29] Thomas Manteufel in *The Bible in the History of the Lutheran Church,* ed. J. Maxfield, 69.
[30] Luther, *Avoiding the Doctrines of Men (1522),* LW 35:139.
[31] A. A. Zaun, "A Study of the Idea of the Verbal Inspiration," 166.
[32] (1529), LW 38:19.
[33] Thesis XVIII. C. F. W. Walther, "The Evangelical Lutheran Church, The True Visible Church of God on Earth," in *Walther and the Church,* eds. Wm. Dallmann, W. H. T. Dau and Th. Engelder (St. Louis: Concordia Publishing House, 1938).

by divine authority, but in no way can human proof for the truth do anything but undermine the teaching of Christ. Since Christ is God, faith in Him is of divine character and certainty. "No extra-Biblical material, philological or historical, may determine the exegesis. . . . All the historical background necessary for the correct understanding of Scripture is given by Scripture itself."[34] God does not misspeak, forget key points, or purposely leave man in the dark when He speaks.

Scripture's words set a hard limit to the boundaries of theology. The words define divine knowledge. "We must avoid seeking beyond that which is necessary and has been written."[35] Doctrine must be divinely grounded, even if tradition would seem solid enough. Man's opinion is uncertain, while God's words bring the Spirit's conviction. "Even if all saintly teachers maintained this or that, it would mean nothing over against a single statement of Scripture."[36] God communicates in words, not in an irrational, indescribable way.

The idea of interpreting "Scripture as [a] whole" was "invented by Schleiermacher" to get away from the offense of certain passages and their burdensome authority.[37] Where the words do not rule, man picks and chooses as he sees fit, as the tyrannical lord of the scriptural manor. Modern commentaries obscure the truth of God's words and sit in critical judgment over them, rather than submit to the truth. If God, who is love, speaks candidly in Scripture, we should think differently: "Would it ever occur to a lover to read a letter from his beloved with a commentary?"[38] So much exegesis of the last few hundred years is a detour, a way to evade placing oneself under the text and listening to God. Pope Leo XIII, citing 2 Tim. 3:16, called the entire Bible God's "own oracles and words—a Letter, written by our heavenly Father."[39]

The charge is often leveled that "the Bible's nature is deduced from

---

[34]F. Pieper, *Christian Dogmatics*, 1:101.

[35]M. Flacius, *How to Understand the Sacred Scriptures*, 96.

[36]Luther (1518), quoted in: J. M. Reu, *Luther and the Scriptures*, 15–16.

[37]Thomas Manteufel in *The Bible in the History of the Lutheran Church*, ed. J. Maxfield, 80.

[38]Sören Kierkegaard, quoted in: *Inerrancy*, ed. N. L. Geisler, 116. Most early Protestant Bibles abstained from exegetical notes. Is the existence of a "study Bible" a subtle confession that the biblical text is unclear and needs extra-biblical material to make it relevant? D. M. Farkasfalvy, *Inspiration and Interpretation*, 159.

[39]"*Providentissimus Deus*: On the Doctrine of the Modernists," 3–4.

a given conception of what a book should be like if it was written by God."[40] It is the reverse that is done in actuality by the modernists. We have a book that speaks sufficiently for itself, but human reason legislates that a divine book cannot exist unless it is proved by man according to the evidence. Any other method would be unscientific. That is why citing Bible verses does not "prove" anything to the modernist.

God is not merely a rational thought of man—His ways are higher than man's. "It has pleased the Lord to say much that seems wrong and impossible" in Scripture.[41] Yet it does give life because it is the Savior's speech and gives knowledge of Him. God will punish those who add or take away from the *words* of the book:

> I warn everyone who hears the words of the prophecy of this book: if anyone adds to them, God will add to him the plagues described in this book, and if anyone takes away from the words of the book of this prophecy, God will take away his share in the tree of life and in the holy city, which are described in this book" (Rev. 22:18–19).

What is the correct method of interpretation? Scripture alone. "Luther's genius is precisely that he did not construct a theology, but dutifully played back what he found in God's inspired Word."[42] His faith in God's Word was proportional to His theological greatness: "If we understood the languages nothing clearer would ever have been spoken than God's word."[43] It is not just a tool to use, but God's mouth to obey and follow. "The principle of philosophy is reason, and [the principle] of theology [is] revelation." "Here beyond question there are two different concepts of truth: the philosophical, which needs compelling proof and is understood; and the theological, which depends only on the authority of Scripture and has to be believed."[44] While the scientific theologian wants objectivity by doubting all, he obtains only subjectivity from his self-centeredness.

Scripture is God's norm, so that it functions in a practical manner.

> In this way the distinction between the Holy Scripture of the Old and New Testaments and all other writings is preserved,

---

[40] T. H. Rehwaldt, "The Other Understanding of the Inspiration Texts," 362–63.

[41] August Pfeiffer (d. 1698), quoted in: R. D. Preus, *Inspiration of Scripture*, 85.

[42] Eugene F. Klug, "Luther and Higher Criticism," *Springfielder* 38:3 (1974), 216.

[43] *To the Councilmen of All Cities in Germany that they Establish and Maintain Christian Schools* (1524), LW 45:364.

[44] K. Scholder, *Birth of Modern Critical Theology*, 122.

> and Holy Scripture alone remains the only judge, rule, and guiding principle, . . . as the only touchstone. . . .[45]

The Scriptures are meant to be used as a simple test to judge all other communication that claims to be Christian. Scripture is a touchstone. "A touchstone, a flint-like stone, sifted genuine from counterfeit gold; when a pure gold object rubbed across it a gold streak showed plainly."[46] It should be used to compare teachings. If its words are not actually used, all the effusive praise of Scripture is worthless. It is not man's estimation of Scripture that matters. The Spirit's physical words rule Christ's Church. When Scripture does not rule verbally, man is allowed the final say.

Scripture alone is not a rational method, a "legalistic exposition of Scripture," or a "legalistic biblicism."[47] Hearing God speak is not a matter of following man-made rules or methods. It is a life-or-death undertaking to hear God speak. The truth leading to eternal life is at stake, so receiving God's Word is never a neutral activity. Perhaps the strongest statement on inspiration in the Lutheran Confessions is connected to the most sublime article of justification by faith:

> Truly, it is amazing that the adversaries are in no way moved by so many passages of Scripture, which clearly ascribe justification to faith, and, indeed, deny it to works. Do they think that the same is repeated so often for no purpose? Do they think that these words fell inconsiderately from the Holy Ghost?[48]

To hear a single word of Scripture is to hear the Spirit, who is true God.

Pre-modern interpreters, following Scripture itself, spoke of suffering and affliction as essential hermeneutical ingredients: "It is good for me that I was afflicted, that I might learn your statutes. The law of your mouth is better to me than thousands of gold and silver pieces" (Ps. 119:71–72). Without external humbling, "reason's proud refusal to be bound by the Word leads only to idolatry"—"to scrutinize the 'essence' (*naturam*) of God, instead of knowing His 'purpose' (*voluntatem*) as it

---

[45] FC Ep Rule, 7; *The Book of Concord: The Confessions of the Evangelical Lutheran Church,* eds. Robert Kolb and Timothy Wengert (Minneapolis: Augsburg Fortress, 2000), 487.

[46] E. F. Klug, "Luther and Higher Criticism," 217.

[47] G. Ebeling, *Word and Faith,* 36, 40.

[48] AP IV:107.

is set before us in Christ."[49] Scripture judges and leads, even when man tries to gain the upper hand. It is "the sole, true liege and lord over all writings and teachings on earth."[50] What is the method of Luther? To bear the cross and listen: "Grief and sorrow teach how to mark the Word. No man understands the Scriptures, unless he be acquainted with the cross."[51]

Moderns set the philosophical idea of Christ the Word against the actual words of Christ in Scripture. But in Scripture, "Christ the Word is a rare and unusual use of the term."[52] This statement, which proves that Jesus is God and Creator, does not conflict with *sola Scriptura:* "In the beginning was the Word, and the Word was with God, and the Word was God" (Jn. 1:1). This description is certainly not a replacement for Scripture as sole judge. When Christ the Word is used against Scripture, the Word is made wordless. "Christ," apart from Scripture, is a mere signifier without definite content. Christ as a pure mental idea is not the object of true faith.

The words of the Gospel, normed by Scripture, deliver the resurrected Christ. Theology is to speak, but what is it to say? Only a single source for theology can lead to a single, definite, ruling theology. Scripture alone subjugates reason and allows no room for it to roam in divine matters. Bible passages explicitly establish doctrine that is more certain than heaven and earth. Fixed, unyielding passages put reason in a passive role. Such "proof-texting" is most hated by moderns because it allows no room for reason to be divine. "Reason does not understand that the most perfect worship is to hear God's voice and believe."[53] As God's mouth, Scripture "in itself is completely certain, clear, open, its own interpreter, testing, judging, and clarifying all things."[54]

---

[49] B. A. Gerrish, *Grace and Reason*, 77.

[50] Luther, quoted in: Hans-Werner Gensichen, *We Condemn: How Luther and 16th-Century Lutheranism Condemned False Doctrine,* trans. Herbert J. A. Bouman (St. Louis: Concordia Publishing House, 1967), 45.

[51] Martin Luther, *The Table Talk of Martin Luther,* trans. William Hazlitt (Philadelphia: The Lutheran Publication Society, 1873; http://cat.xula.edu/tpr/works/tabletalk/), LIV.

[52] S. Becker, *The Scriptures—Inspired of God*, 22.

[53] B. A. Gerrish, *Grace and Reason*, 79.

[54] Luther, *Assertion of All Articles,* quoted in: H. Gensichen, *We Condemn*, 37.

Part VII

Consequences for Theology

Chapter 25

Replacements for *Sola Scriptura*

Divine doctrine is taken from plain statements of Scripture and organized as it is presented by the Holy Spirit, or man pillages the divine Word, justifying and critiquing what is present to find suitable materials to erect his own personal cathedral. This theological system-building is not to the glory of God. It will always be man-centered, though it might seem more spiritual to creatively remake Scripture into a rationally consistent system. However, the scriptural Christ will be lost in such a mental scheme.

Only the bare words of God have divine authority in all matters. To treat the divine words as less than divine because they look human is to deny the only contact point God has given. Christ ascended and will not descend until the Last day. "I know nothing of Jesus Christ but only his name; I have not heard or seen him corporally, yet I have, God be praised, learned so much out of the Scriptures, that I am well and thoroughly satisfied; therefore I desire neither to see nor to hear him in the body."[1] Christ would have all who submit to Him submit to His words. "What does not really and fully norm is no norm" at all, but needs a replacement.[2] To define Christ apart from Scripture is to treat Christ as nothing but a mental idol, moldable to our whims.

---

[1] *The Table Talk of Martin Luther* (1873), CCXXXII [232].
[2] Horace Hummel in *Studies in Lutheran Hermeneutics*, ed. J. Reumann, 217.

The idol-Christ is shaped by reactions against what seems unworthy of Christ to man's reason. This Christ is not the scriptural Christ, but one that reason invents for itself. The scriptural Christ would have all speculation measured by Scripture's words, so that everything is subjugated to Him: "Get behind me, Satan! You are a hindrance to me. For you are not setting your mind on the things of God, but on the things of man" (Mt. 16:23). A Christ who does not speak in Scripture is not the resurrected Christ.

In modernism "the norm of dogma [has] fallen."[3] In pre-modern Christian orthodoxy, "every doctrine of Holy Scripture is set forth at some place or other very clearly, as the main theme of discourse."[4] Theological degrees and scholarly tools are not required to hear God speak, though, without an actual divine norm, man is trapped within his own sinful thinking. The result: "dogmatics has the task of bringing the church's teaching into contact and discussion with contemporary principles of thought, there to submit it to critical sifting and present it in its full inner coherence."[5] The "church's teaching," as a man-made, historical by-product, is quickly replaced by other man-made, more modern thoughts. The great tragedy is that Christians have largely done this to themselves. It is within the Church, by her supposed teachers, that simple Christians are led astray.

God has promised no means of truth other than His written Word. When scriptural authority is denied, the speculation of man reigns. Systematic Christian rationalists "lay down certain principles, called axioms, or first truths of reason, and from them deduce the doctrines of religion by a course of argument as rigid and remorseless as that of Euclid."[6] So the facts of religion are not discovered in revelation, but produced and authorized by the orderly mind. Rationalism, the craving for the impression of coherency within the human subject, has filtered down to even the most uneducated critic. As the saying goes, "everyone's a critic." Bare scriptural proof, to the mind which only admits scientific knowledge, can only be a harmful, alien influence. Divine scriptural words set definite limits to rational consideration, compared to the

---

[3]K. Scholder, *Birth of Modern Critical Theology*, 41.
[4]L. Fuerbringer, *Theological Hermeneutics: An Outline for the Classroom* (St. Louis: Concordia Publishing House, 1924), 16.
[5]G. Ebeling, *Word and Faith*, 27.
[6]Charles Hodge (d. 1878), J. C. Livingston, *Modern Christian Thought*, 306.

authority-denying freedom of historical criticism.

Many modern theologians are preoccupied with Martin Luther's theology, but study it from a scientific perspective. Scholarly Luther research cannot attain the fullness of Luther's teaching, because it denies the only authority to which Luther held. It has its own academic, critical standards at odds with Luther's biblical submission. The search for themes or motifs, while appealing to modern minds, is entirely man-centered. It is an academic Easter egg hunt for the secret key to unlock the whole. "The temptation to seek one focal point or one foundation[al] theme in terms of which everything else can be explained is pervasive in Luther research. Such a temptation is in itself a tribute to the depth of Luther's theology, but it must be resisted."[7] Luther was not a system-producing rationalist. The key to the complexities of Luther's thought is God Himself. Binding oneself to the lowly book of Scripture is the only infallible principle and means of knowing God's will. This is a spiritual endeavor, requiring the Holy Spirit given in Baptism, not academic degrees. Historical-critical distance and academic objectivity will never allow a proper confession of truth itself, so Luther's teaching will puzzle scholars who try to get above it. Luther's theology cannot be apprehended scientifically or thematically. Its power is from Scripture's divine words. Treating divine doctrine as man-made means that God is not allowed to be true God. Luther did not even consider himself above the Small Catechism, though he wrote it. Following him, "Walther based his doctrine on *nuda Scriptura*," the plain words of Scripture.[8]

The modern shift in understanding Scripture has led to a heavier emphasis on outward things, at the expense of God's Word. "Divine liturgy" or "divine service" became the high church replacement for the less holy Scripture.[9] The idea of a spiritual continuity guaranteed by the repetition of external activities of sinful participants is essentially enthusiasm. Service orders are all man-made, even if the content is entirely biblical. No one order of service has divine priority—"liturgy"

---

[7] Jonathan D. Trigg, *Baptism in the Theology of Martin Luther* (Leiden: Brill, 2001), 2.

[8] T. Engelder, *Haec Dixit Dominus*, 26.

[9] Luther can only call "divine service" faith itself, not an assemblage of externals practiced weekly. True worship, according to Christ, is not at a particular location or time, but in the Spirit and truth (Jn. 4:23). Without the Spirit and faith there is no worship, no matter what man does.

is not a magical word that makes the Spirit present. A selective use of the scriptural words, ingrained by tradition, is not somehow more holy than the source and context of those words—Scripture itself. It is the proclamation normed by Scripture that makes Christians and preserves the true sacraments, not mere external sameness and man-centered continuity. "Nothing can be affirmed of God's will without God's Word."[10]

Scripture itself has few practical requirements for public worship, though doctrinally it has much to say about why order should exist in the outward service. To both high-church and low-church enthusiasts, what people do in the service is more important than the Word of God preached. Activity takes the place of doctrine for moderns, who see submission to Scripture as evil: "Liturgy is an essential hermeneutical link between the past and the present."[11] Outward things replace the spiritual power of the Word. Here is the root of the worship wars: competing replacements for the divine authority of God's Word. The modern accentuation of either old or contemporary rituals is rightly seen as rooted in the same dissatisfaction with the plain Word of God. Bare human activity, even in the best ecclesiastical acts, can never be a replacement for the Spirit working through the external Word. Externals, even the divinely instituted ones, are not a replacement for certain doctrine—the revelation of God's will.

Liturgical theology is a modern replacement for biblical theology: "the values of liturgical renewal . . . echo the shifts in revelation theology."[12] Liturgy is posited as one continuous, adapting, ongoing revelation, so the Spirit's actions can be recognized externally. The "temptation to describe worship in romantic terms," instead of sinners face-to-face with God's justifying Word, is strong for modernists.[13] The outward actions of man, rather than trust in the word-action of the Spirit, are given first place where the Word of God is distanced from people.

Luther spoke of man-made orders of service in law terms: "The inexperienced and perverse youth need to be restrained and trained by

---

[10] AP XV:17.

[11] C. E. Braaten, *History and Hermeneutics*, 158.

[12] Philip Caldwell, *Liturgy as Revelation: Re-Sourcing a Theme in Twentieth-Century Catholic Theology* (Minneapolis: Fortress Press, 2014), 104.

[13] Walter Sundberg, "Wilhelm Löhe on Pastoral Office and Liturgy," *Word & World* 24:2 (2004), 196.

the iron bars of ceremonies," so that God's Word will not be impeded.[14] All ceremonies, except the ones instituted by Christ in Scripture, are optional to God, while true worship is of the Spirit and faith (Jn. 4:23). Without the Spirit all churchly ceremonies, even the divine sacraments, become unprofitable idolatry. The Lutheran Confessions grant Christ's body and blood in the Roman Supper, but because of their doctrine, they call the Roman mass false "Baal *Gottesdienst* [worship]."[15]

The liturgical renewal movement does not distinguish between what is man-made and optional and what is divinely-commanded. It is based upon a modern focus on efficacy, without clearly defining the nature of things according to God's Word. Luther's teaching contrasts sharply with this vague romanticism: "if someone tries to besmear [Scripture] with his reason or your own thoughts, then answer: Here I have the plain word of God and my faith; to this will I hold without further thinking, questioning, listening, nor asking whether it harmonizes or not . . . ."[16] Liturgical theology uses a still active tradition, but without any divine authority. Man's thoughts rule when he tries to fill the void in the wake of the denial of the scriptural text.

The Lord's Supper becomes a meaningful liturgical activity and the center of the service without having to define what is received by unbelievers. Faith becomes indefinite, and liturgical words are held in mystical devotion because of man's use of them. As un-theological as the Bible may seem to critics, every replacement for plain Scripture is man-centered, and every church that loses its footing on Scripture loses the Gospel eventually. Only God's Word, not the elevated traditions of man, can sustain faith. Man's activities are poor replacements for God's authoritative speech.

Sola Scriptura "means at its root 'Scripture interprets itself' and 'Scripture authorizes itself.'"[17] This does not mean that any use of Scripture will be correct or Christian. Many misuse it or carry to it their own ideas. Even strong believers in inspiration have imbibed the

---

[14] *The Freedom of a Christian* (1520), LW 31:375.

[15] Ap XXIV, 98.

[16] J. M. Reu, *Luther and the Scriptures*, 82.

[17] David W. Lotz, "Luther and *Sola Scriptura*," in *And Every Tongue Confess: Essays in Honor of Norman Nagel on the Occasion of His Sixty-fifth Birthday*, eds. Gerald S. Krispin and Jon D. Vieker (Dearborn, MI: The Nagel Festschrift Committee, 1990), 256.

modernist spirit and lord themselves over God's Word. Because of sin, the temptation always exists to humble the Word of God, instead of ourselves. But the "alone" of "Scripture alone" means that every scriptural teaching and conclusion is from God, not man. The denial of Scripture's authority leads to every man's reason possessing prophetic status: "It is foolish to ascribe inspiration to special persons; everyone ought to be a Christ, a prophet, an inspired man." This "new atheism" is not so much a logical rejection of the deity, but an elevating of man to divine status. This method of divining knowledge has become a right, and apart from it truth is considered absent.[18]

Respectability adheres to the university model of theological education. To teach God's Word, a graduate degree is needed more than the fruits of the Spirit. The content of academic discourse is not normed by Scripture but by prevailing methodology. Only atheistic objectivity is scholarly. "Evangelical graduate students have sometimes sold their souls for a Ph.D. degree."[19] Many learn critical methodology from pagan professors to be professors themselves. But method quickly becomes a way of life—a golden calf to which one must sacrifice everything. "Every student who entrusts himself to the university must accept the yoke of the atheistic intellectual starting point as an inescapable necessity."[20] One may not start with the whole truth of God. To do so would be unscientific, and therefore unscholarly.

Cartesian theologians "take the consciousness as their starting-point," no matter how traditional the conclusions may appear.[21] In contrast to the scriptural approach, which does not go beyond the Spirit's words, it is claimed that without a Christological or "trinitarian perspective no doctrine can be considered fully presented."[22] This is axiomatic theology, where truths are made to fit into a rational system. Theology is made into an intellectual jigsaw so that "everything is deduced from a central truth and all parts form a harmonious whole."[23] The basis of apparent scriptural teaching is man, since it is authorized by him. Whatever

---

[18] *Library of Universal Knowledge*, 621, 622.
[19] J. Barton Payne in *Inerrancy*, ed. N. L. Geisler, 108.
[20] E. Linnemann, *Historical Criticism of the Bible*, 33.
[21] H. Thielicke, *Prolegomena*, 115.
[22] David P. Scaer, "Biblical Inspiration in Trinitarian Perspective," *Pro Ecclesia* 14:2 (2005), 1.
[23] T. Engelder, *Reason or Revelation*, 57.

does not fit man's system built on the whole of Scripture is thrown overboard. But Jesus referred to single texts, not even His own inner perspective—which is a powerful argument against Cartesian doubt. However neatly man's thoughts fit into the Christian tradition, the temptation to go beyond the words of the Bible is satanic. It attempts to do theology in a better way than Jesus, the true God.

Systems, even when based on biblical ideas, are unhinged when Scripture does not measure all theological speech. Even law and Gospel, when used as independent principles, are un-Lutheran. Hence the modern denial of the third use of the law[24]—man wants to define the law against Scripture. Yet in Scripture, the law does not lose its inherent validity when the Gospel is proclaimed. The Bible presents a more nuanced view than the rationalization that the law is a negative force canceled by the greater positive force of the Gospel. Christ, the guilt-bearer for all mankind, is lost in this scientific description of supposedly opposing forces. "Nothing should be asserted in [questions of] faith without scriptural precedent."[25] For the one starting with critical principles, no single passage of Scripture will bear the full weight of any particular conclusion.

Law and Gospel are not ideas to be flippantly manipulated and philosophized about, but truths bigger than all men that kill and make alive. Christ is the center of Scripture, but the doctrines of law and Gospel are determined by Scripture itself. This explains why we have sermons *about* law and Gospel and their effects, but precious little divine law and divine gospel bearing down on man. Moderns speculate critically about the hypotheses of doctrines and themes, but apart from Scripture no man can speak with divine authority. "The slogan 'sinner and righteous at the same time' does not float in the air, or stand by itself an axiom from which conclusions can be deduced."[26] One must speak with divine authority, as God speaks in Scripture, to rightly proclaim God's Word.

Critical uncertainty does not allow speaking to another on behalf

---

[24]This is the informational or rational use that says the Law is more than a imaginary, condemning power. The law (as God's will) has a content that is useful, valid, true, and holy, even for the believer who is forgiven and no longer under the law and its curse (Rom. 7:12).

[25]Luther, *Against Latomus* (1521), LW 32:230.

[26]David S. Yeago in *Rightly Handling the Word of Truth,* ed. C. E. Braaten, 68.

of God, with His full authority. This error strikes at the heart of Christianity, because the pastor with God's Word does speak for God. "And because God truly quickens through the Word, the keys truly remit sins before God. . . . Wherefore the voice of the one absolving must be believed not otherwise than we would believe a voice from heaven."[27] Note that this is not referring to a specific rite or formula, but following Scripture it describes the general *power* of the keys. No specific liturgical formula of absolution is scripturally authorized. A pastor must preach God's Word in the stead of Christ, but without Scripture he merely plays liturgical dress-up or is a non-confrontational narrative-teller. The called man of God speaks for God insofar as he remains a servant of His Word, relying on the words given by the Spirit.

Law and Gospel have become tools used against God, entirely man-made ideas, not different words of the same God. "To say we must bow down to what God said, no matter what he has said is to blur the distinction between Law and Gospel."[28] But is not the law the will of God, even when it means death to the sinner? "For the Law is a mirror in which the will of God, and what pleases Him, are exactly portrayed, and which should be constantly held up to the believers and be diligently urged upon them without ceasing."[29] Law and Gospel are from God, and defined exclusively by Him in Scripture. Man's imagination does not have authority to fill the content of these words. According to one critic, "the *sola fide* [teaching of faith alone] destroys all secretly docetic views of revelation which evades the historicalness of revelation, . . . a sacred area from which the critical historical method must be anxiously debarred." This modernist "faith" results in "the critical destruction of all supposed assurances."[30] Here historical criticism is identified with the Gospel itself, so that man, by critical faith, denies Scripture and all authorities, thus becoming a god. In scientific analysis God is not allowed to speak without man first having the right of critical refusal. It is not Scripture that has become an idol, but the man-centered method promising an anarchical freedom.

Two examples of popular, rationalizing systems are Luther's "two

---

[27] AP XII:40–41.
[28] A Concordia Seminary, St. Louis professor (1970–71), quoted in: P. A. Zimmerman, *A Seminary in Crisis*, 177.
[29] FC SD VI:4.
[30] G. Ebeling, *Word and Faith*, 56, 59

kinds of righteousness" and his "theology of the cross." But they are merely trite descriptions. Luther did not do theology on the basis of axioms. Yet, with much fervor, entire theologies are erected on the basis of a few words of a sinner. It is quite telling that Lutheran scholars latch on to various, conflicting "theologies of the cross," while Luther quickly stopped using that terminology. The search for themes from which to make judgments is pure philosophy.

> So what is the meaning of this cry: "The old inspiration doctrine is nothing, it does not do justice to the human side of Scripture"? What do they want to say? They want to charge that we err in not being willing to recognize any errors in Scripture, but regard it not only as the best book, but as the book of truth in the highest and fullest sense of the word. They want us to join them in denying the divine origin of the book; they want us, as often as we have a passage before us, to say with that spirit: "Has God really said this?"[31]

"The loss of *sola Scriptura* only leads to a new sacerdotalist (church-matrix of tradition), new clericalism (expert-gnosis) and mystical agnosticism."[32] The sacraments are paramount in high-church romanticism, but if Scripture is not the Word of God, how can people know if the sacraments are instituted by Christ? Who defines them? The sacraments may bring a "Christ," but doctrine establishes which Christ is being relied upon. The holy means of grace can easily be praised as human traditions and justifying works in their mere performance. This works righteousness must eventually occur apart from scriptural preaching and teaching. The preaching of faith allows the sacraments to be promises of Christ and forms of the one, true Gospel. To be correctly trusted in, they must be divinely authorized.

Luther rightly proves the nature and benefits of the sacraments by citing single texts of Scripture. An abundance of flowery, human words is no substitute for a certain and divine Word. Ceremonial infatuation sees visible things as more real than the power of God's Word, but the Spirit who regenerates only comes through the Word of Christ. Without definite, instituting words of God, the sacraments are not *divine* sacraments, lacking the warrant to be trusted for divine forgiveness. They can easily be idols when viewed as justifying works apart from fear,

---

[31] C. F. W. Walther, "Evening Lectures on Inspiration," Lecture IV.
[32] C. H. Pinnock, *A Defense of Biblical Infallibility*, 7.

love, and trust in the true God: Christ Jesus. Sinners are always induced to hold up what they do, especially if the act is divinely instituted, as righteous works which make the true Christ unnecessary. Talking about the sacraments is no replacement for the doctrinally established scriptural Christ.

When the Gospel becomes intangible and disconnected from divine words, earthly things naturally take its place. "Single passages of Scripture are frequently used as isolated truisms to support ideas that have little to do with the Gospel."[33] This unscriptural Gospel becomes such a minimal gospel that it actually has no content at all. Man uses the jackhammer of the word 'Gospel' to critically rearrange Christianity according to his own reason. Truisms like the Ten Commandments or Luther's Table of Duties (the direct application of Scripture passages to present-day vocations) become hindrances to a man-centered Gospel. A holy Gospel "vocation" can be defined as widely as ordering a $7 latte, if God does not define scripturally what vocations are godly. The Gospel simply becomes the critical principle which destroys all knowledge. But the scriptural Christ never warns against taking Scripture too seriously—He submits Himself to the plain text. Why are moderns too proud to follow our Lord and take up the cross of being called a fool?

Receiving human words as divine, such as those spoken through a parent or pastor, is a basic tenet of Christianity: "And we also thank God constantly for this, that when you received the word of God, which you heard from us, you accepted it not as the word of men but as what it really is, the word of God, which is at work in you believers" (1 Th. 2:13). The content of Scripture reproduced by sinful pastors is still the almighty Word of God, in a derivative sense. It has the same efficacy and spiritual power as Scripture, though it cannot establish new articles of faith. This is theology from above, from God Himself. All theology apart from the given words is man-centered, a theology from below, a "theologizing from the Christian consciousness."[34] Luther's dictum is quite clear-cut:

> With anyone who denies that the evangelical Scriptures are the Word of God I will not argue a single word; for we should not enter into dispute with anyone who rejects the

---

[33] P. H. Nafzger, *These Are Written*, 151.

[34] A. Hoenecke, *Evangelical Lutheran Dogmatics*, 1:477.

first principles (*prima principia,* perhaps in the sense of 'primary sources').[35]

Do we trust the word of reputable scholars more than the Word of God? There is a current need for more judging—instead of a naïve trusting in what sinners say. The spiritual man judges all things, but does so by God's own critical authority—His very words (1 Cor. 2:15). Where Scripture does not reign, God does not—and man is lost.

Reason, when it judges according to its own standards, remains sinful, even for the Christian. "Whatever reason brings out of its stinking heaps must be subordinate to the Word of God and be measured by it as its touchstone . . . ."[36] The doctrine of sin is closely connected to one's view of Christ's Scripture. The counterpoint to sin is God's true, vivifying Word. It does all the work of making Christians and giving Christian knowledge. Through the Word the Spirit gives divine conviction, so that "you will not be bold or very confident without Scripture."[37]

Church

The Church is said to speak for God when Scripture is silenced, but theologically the Church is only passive. She does not do or speak in relation to her Head. She is washed, cleansed, and presented holy by Christ (Eph. 5:25–27). Sinners, even if forgiven, do not determine truth. The Church consists of those who receive true knowledge of Christ and are forgiven by grace.

The Church as the locus of revelation, or mouthpiece for God, is enthusiasm. It looks not to the pure Church of Christ which is composed of sinners righteous *in* Christ, but the opinions of sinners. No amount of forgiveness allows one to speak for Christ. Men conflict with each other and Scripture, therefore they must be ruled by Scripture. The Church only hears in relation to Christ. Believers are justified passively by the speaking Christ.

---

[35] Quoted in: J. Köstlin, *The Theology of Luther*, 2:227.

[36] Nicolaus Arnoldus, quoted in: K. Scholder, *Birth of Modern Critical Theology*, 119.

[37] Luther (1516), quoted in: Mark Thompson, *A Sure Ground on Which to Stand: The Relation of Authority and Interpretive Method in Luther's Approach to Scripture* (Waynesboro, GA: Paternoster, 2005), 7.

Modern Protestants emphasize the Church more than pre-modern papists.

> Inspiration therefore must not attach to a small number of exceptional persons like St. Matthew or St. Paul: it must extend over a large number of anonymous persons—In this sense scripture emerged from the community: it was a product of the church.[38]

The loud majority suffices for visible proof of the Spirit, providing the desired scientific evidence. But the true Church is invisible to man, an article of faith. Outwardly it is recognized only by Christ's teaching. She is not the norm of her own teaching: "The church as one communal prophet living across time cannot fatally err, not if the Gospel itself is true."[39] The question is whether a particular group's confession has anything to do with the real Christ, who is the Head of the Church. Without Scripture it is impossible to distinguish between competing spokesmen for Jesus and His one Gospel.

What is meant today by "the Church"? Often the largest vocal majority, led by the magisterium of exegetical scholars. Atheistic methods lead critics to want to be what they cannot find: a divine revelation. Authority assumed by human leaders apart from scriptural mandate is political. Without the words of Scripture, even the divinely established pastoral office becomes a tool for secular power.

The authority of any office is limited by God's design and bound to His institution. The only spiritual power is the Word. A minister has no intrinsic authority in place of a supposedly absent Christ. Judas was called and had the apostolic office, but it did not prevent him from betraying Jesus. Even after the giving of the Spirit, the apostles did not become perfect men. Their teaching, in written and oral form, established the Church, because they spoke Christ's own Word. Yet the oral aspect of their office is inaccessible today. Only by virtue of the Spirit's written Word is that historical distance bridged. "Scripture itself makes no distinction between what the holy writers once preached orally and what they wrote."[40]

A hermeneutic of doctrinal development allowed modern Lutherans,

---

[38] James Barr, quoted in: M. Ruokanen, *Doctrina divinitus inspirata*, 145.

[39] Robert W. Jenson, *On the Inspiration of Scripture* (Delphi, NY: ALPB Books, 2012), 44.

[40] Thesis I.3.a. C. F. W. Walther, "Evening Lectures on Inspiration," Lecture III.

like Wilhelm Löhe, to question "those aspects of Reformation theology they found unsympathetic to their conservative cause. For example, Luther's early commitment to the priesthood of all believers" and his rejection of the divine institution of ordination were viewed as needing to be modernized. Specifically, changing political circumstances and an intense focus on the office of the ministry motivated these modern theologians to go beyond the teachings of Luther and the simple words of the Confessions:

> The fact that the Augsburg Confession grounds ministry in the modest idea of 'a regular call' (article 14) without specifying any particular rite or ceremony for ordination was taken as proof that the Reformation's doctrine of ministry stood in need of theological reappraisal.[41]

Scripture, however, lacks any direct, divine institution of this ceremony. Instead, the loss of scriptural authority led to an increased emphasis on ordination and going beyond Luther's plain scriptural thinking: "I personally find it very difficult to designate as a human rite or adiaphoron any ceremony in which God is the Giver and the Holy Spirit is the recipient."[42] Either God's Word establishes the rite and makes it divine, or Christ does not institute it in Scripture, leaving it with only human authority. The void of divine authority, born when Scripture is deemed insufficient, leads to deformities in all areas of doctrine. Man will use his reason to sand down the troubling spots and fill in the gaps with baseless opinions.

That ordination was a valued apostolic practice is the most one can say about it scripturally. It is mentioned in passing, but not commanded. Following good order and apostolic custom, it is therefore not to be discarded, but it cannot be the basis for understanding the ministry of preaching. No scriptural text can bear that weight. Following Luther, Walther correctly confessed: "Ordination is not of divine institution but is an apostolic ecclesiastical arrangement and only a solemn public confirmation of the call."[43] This is offensive to many modern, conservative

---

[41] W. Sundberg, "Wilhelm Löhe on Pastoral Office and Liturgy," 193.

[42] This does not happen apart from God's Word or Baptism. Pastors do not receive a different Spirit from the one received by all at Baptism. David P. Scaer, *Ordination: Human Rite or Divine Ordinance* (Fort Wayne, IN: Concordia Theological Seminary Press; http://www.ctsfw.net/media/pdfs/ScaerOrdinationHumanRiteorDivineOrdinance.pdf), 12.

[43] Concerning the Holy Ministry: Thesis VI.B. C. F. W. Walther, *Walther on the*

Lutherans. Might they want some personal authority to speak for a silent, voiceless Christ? Scripture remains the only theological authority for doctrine. The urge to find a more rationally coherent picture than the plain scriptural words is the satanic spirit of modernism at work: "Is it possible to expect that we shall make a sacrifice of understanding, *sacrificium intellectus,* in order to accept what we cannot sincerely consider true—merely because such conceptions are suggested by the Bible?"[44] When Christ speaks, the answer is "yes."

The appeal to sinful theologians, even ones enlightened by the Spirit, is a sad replacement for the divine revelation of Scripture. "Divine inspiration is the divine decision continually made in the life of the Church and in the life of its members."[45] If outward churches actually believed the same, this would at least make sense. Rather, this new approach of continuing revelation seemingly validates everything. The ecumenical spirit, which desires visible unity at any cost, accepts all words, except authoritative ones which cause divisions. It appears to be a direct outgrowth of the modern, atheistic spirit—it wants scientific evidence of oneness. God works to get results for moderns, but does not speak, leaving no rational basis for theological statements.

Without the stable meaning of Scripture, fellowship is at most a dicey proposition. The Church becomes not sinners gathered around God's Word, but the impure source of partial truth. True fellowship in Christ requires more than man's sins—it must be based on Christ and His words. The ecumenical spirit is results-oriented rather than doctrinally-oriented. Instead of proclaiming, it divinates the truth from popular culture, bowing to man's words. The unimpressive results of God's humble Word are assaulted with critiques.

The Church in her outward actions and institutional decisions is simply a mass of sinners. Her holiness is hidden in Jesus. The justified believe in the truth which comes from Christ; they do not fashion the truth in the workshop of their own evil minds. Sinners outwardly joined together are not a divine authority for themselves. "We do not concede

---

Church: Selected Writings of C. F. W. Walther, trans. John M. Drickamer (St. Louis: Concordia Publishing House, 1981).

[44] Rudolf Bultmann, *Jesus Christ and Mythology* (Upper Saddle River, NJ: Prentice Hall, 1958), 17.

[45] Karl Barth, *Church Dogmatics,* quoted in: S. J. Grenz and R. E. Olson, *20th-Century Theology,* 71.

to them [the papal bishops] that they are the Church, and indeed they are not; nor will we listen to those things which, under the name of Church, they enjoin and forbid. . . . The Church is, namely, the saints, believers, and lambs, who hear the voice of their Shepherd."[46] The Church is the sum of Christ's sheep. They are justified by the absolving voice of Christ. The Church is defined by God's Word, His very voice, not vice versa.

---

[46] FC SD X:19.

Chapter 26

Theological Chaos

The wise men of the age, in the name of God, use their personal gospels to go beyond Scripture.

> Although they knew God, they did not honor him as God or give thanks to him, but they became futile in their thinking, and their foolish hearts were darkened. Claiming to be wise, they became fools, and exchanged the glory of the immortal God for images resembling mortal man (Rom. 1:21–23).

While the pure Gospel is simple, when Scripture's words are denied it is lost. All that is left is Christian vocabulary without any divine substance. "The doctrine of justification [has] already been in danger of becoming simply God's yes and nothing more, or a mere 'being accepted' and nothing more!"[1]

Christian-sounding words, without scriptural backing, lack real substance and conviction. Sinners are left with a cotton candy gospel that cannot sustain them in temptation and death. Luther makes the cogent point that the greatest confession of faith is submitting to martyrdom *for written words*. Who would "shed their blood for something uncertain and obscure?"[2]

Luther, who confessed that every scriptural letter is from the Spirit, believed that "there is not a superfluous letter in the Scriptures."[3] But with only a human origin, they are all superfluous to modern man. There

---
[1] G. Maier, *End of the Historical-Critical Method*, 17.
[2] *The Bondage of the Will* (1525), LW 33:100–101.
[3] *The Three Symbols of Creeds of the Christian Faith* (1538), LW 34:227.

are frightening statements made by many so-called "theologians"—those who claim to speak for God. "The objections to verbal inspiration . . . serve as a shining example of how God inflicts His just judgment upon all critics of His Word—they lose their common sense and become utterly unreasonable and illogical."[4] Man's opinions become the authority to which Scripture must become a footnote. Despite criticism abounding, the result is as interesting and appealing as blanched cardboard.

> A *too* scrupulous fear of going counter to established opinion (which fear he conceives to be a natural result of much, and the deepest reading) tends to restrain independent thought; and leads insensibly to the error of identifying Scripture itself with human interpretations of it.[5]

To be modern is to be atheistic—completely unwilling to submit to God's written authority. Methodology offers the promise of rational salvation to the scholar. It is no surprise that many nominal Christians are practical atheists, who live as if there is no God, without any fear of contradicting Scripture. Without an actual divine revelation, preaching is simply man's words, easily ignored. Without God's literal words, how can one stand above another to condemn open, public sins and flagrant ungodliness? It will be one man's opinion against another sinner's word. Forgiveness without repentance is merely a word game, a saltless universalism: "Christ died so I could go on sinning, thanks be to sin." This is the result of a propositionless, contentless Christianity.

Critical scholars are like children at the playground. Few actually criticize the prevailing fads. Ironically, they do not investigate in a truly scientific and independent fashion. The "papacy of scholars" rules over men instead of Christ.[6] The "hermeneutic problem proves to be of fundamental significance" for theology, because it assumes and demands a modernizing of the Gospel—an attempt to translate the Gospel for an atheistic people.[7] Recent history is replete with failed attempts to come up with a better gospel than the one Christ gave in Scripture.

"The historical interpretation has made it clear that there are more or less diverse, if not contradictory, theologies in the Bible itself."[8]

---

[4] F. Pieper, *What is Christianity?*, 243.
[5] J. Miller, *The Divine Authority of Holy Scripture Asserted*, iv.
[6] G. Maier, *Biblical Hermeneutics*, 167.
[7] G. Ebeling, *Word and Faith*, 28.
[8] Duane Priebe in *Studies in Lutheran Hermeneutics,* ed. J. Reumann, 307.

Multiple theologies must lead to multiple faiths, multiple justifications, and multiple truths, if faith, justification, and truth are concerns for the modern at all. Each biblical book or phrase can be a unique theology, leading to the un-Christian language of: "*a* theology of . . . ." If there is not a single teaching of Christ, there is nothing definite to believe. The holy Gospel is perverted into a "form of self-deification."[9]

When no revelation is believed or accepted, every person assumes the place of God. But God is one and speaks with one voice: "Long ago, at many times and in many ways, God spoke to our fathers by the prophets, but in these last days he has spoken to us by his Son" (Heb. 1:1–2). It is God the Spirit who speaks in Scripture. He does not submit to rational proof—just as Christ did not appease scoffers while on the cross. So the diversity evident in Scripture does not bother the believer. Faith and critical doubt are as far apart as the east is from the west.

Moderns are left only with the theological verbiage of very smart men. Despite all the trappings of Christianity, there is no truth. The foundation of divine knowledge has been denied. Factual and divine assertions cannot be made because there is no basis in the minds of moderns immersed in methods of doubt. "Theological" has become code for "harmless fiction." Theologians have big words, but use them to say nothing. Without Scripture no one can speak with authority, as Jesus did. Even our God did not consider it beneath Himself to quote Scripture, referring to the minutest elements of syntax.

---

[9]T. Engelder, "Verbal Inspiration: A Stumbling Block," Part 15, 824.

Part VIII

Christ's Norm for Theology

Chapter 27

Christ and Scripture

Traditionally, true knowledge of Christ was seen as revealed by God, either from Scripture alone or in combination with revealed tradition. In modern theology "Christ" is an iconoclastic principle used to destroy the possibility of any divine knowledge. For example, Karl Barth had a "highly christocentric view of revelation, . . . identified with Jesus Christ and him alone."[1] This sounds very attractive, but if an idea of Christ is the only revelation, God is inaccessible and wordless. A name without any revealed content is not a revelation in any meaningful sense. In modern theology the Christological axiom—a rational, mental principle—is played against the scriptural Christ.

The biblical question must be: which Christology, or Christ, is the right one? Is the word "Christ" merely a password into a secret, hidden, anti-historical knowledge known only by the enlightened scholar? Luther counters: "God will not give you his Spirit without the external Word."[2] The split between Christ and Scripture, and therefore between God and man, is at the heart of modernist errors.

Barth began with God's complete transcendence, so that "it would compromise the sovereignty of God if He were involved in the worldly process."[3] "The critical approach . . . has always sought to drive a wedge between 'the divine and the human' in Scripture, the heavenly and the

---

[1] Avery Dulles in *The Authoritative Word: Essays on The Nature of Scripture,* ed. Donald K. McKim (Eugene, OR: Wipf and Stock, 1983), 240.
[2] *Preface to German Writings* (1539), LW 34:286.
[3] H. D. McDonald, *Theories of Revelation,* 1:73.

historical."[4] So tolerating this method means a total reorientation of theology. The vocable "Christ" becomes paramount to critics, but it is used in an atheistic sense so that it says absolutely nothing. Instead, the scriptural truth of Christ demands the confession not just of verbal inspiration but that God does not come to man non-verbally. He comes *only* in words. To have God's Word is to have God the Spirit who leads to Christ.

The approach of grounding religious authority, not in inspiration, but solely in the person of Christ is thoroughly modern. It dates at least to the early 1800s. Samuel Taylor Coleridge said that "in establishing [the Bible's] authority we must go to the Lord Jesus Christ and make the doctrine of Scripture Christo-centric," meaning faith gives authority to Scripture.[5] It is telling that doctrine is "made" and fashioned by man in accordance with some supposed higher truth. However, "it is not Christianity that needs to be made reasonable. It is reason that needs to be made Christian."[6] Fleshly, system-building reason is put to death through suffering and despair.

Modern theologians "view . . . the Word of God [as] centered in Jesus Christ rather than a book," so they can be oracles for God.[7] Uplifting Christ sounds very Lutheran and Christian, but this rationalist thought pits Christ against Christ's words. This "false antithesis" is based on the assumed incompatibility of God and man.[8] It is also modern idolatry—Christ becomes the prime philosophical axiom, the pivot point of all reality: "It is only through Jesus Christ that history can be defined as world history," or, according to Barth: "In the history of Jesus we have to do with the reality which underlies all other reality."[9] This sounds Christian and is replete with the traditional vocabulary, but it is in reality atheistic philosophy.

"Christomonism is a term that has been used to describe the virtual

---

[4] *Studies in Lutheran Hermeneutics,* ed. J. Reumann, 43.
[5] H. D. McDonald, *Theories of Revelation,* 2:299.
[6] S. W. Becker, *The Foolishness of God,* 159.
[7] *Studies in Lutheran Hermeneutics,* ed. J. Reumann, 62.
[8] Henry J. Eggold, "A Man's View of the Word of God Determines His Attitude toward Inerrancy," Faculty study paper (Copy in Concordia Theological Seminary Library, Fort Wayne, IN), 9.
[9] E. Brunner, *Revelation and Reason,* 405; Quoted in: Matthew Rose, "Karl Barth's Failure," *First Things* 244 (2014), 42.

separation of the person and work of Jesus of Nazareth from God the Father and God the Spirit."[10] Christ did not take on all human words, all human flesh, or all reality, but one human body in an eternal union. Monistic theology "refers all phenomena of the spiritual and physical worlds to a single [rational] principle."[11] In Barth's scheme, "the truth is God's revelation in Christ, and therefore dogmatics is essentially Christology."[12] This Christology, though, is a man-made, rational axiom, since the historical Christ is unknowable by critical means.

The words Scripture uses are not forever united to the divine nature. Rather, very specific words were given by the Spirit to be written for our salvation. Scripture is simply God's speech, which is the real import of inspiration. "The Bible is a product of a two-fold authorship—God's and man's, so it is all divine and all human."[13] Divine authorship is what sets Scripture apart, but the words of the Bible are not Christ's flesh. Where the Spirit is denied interaction with man through words, Christ is in grave danger of becoming a metaphysical principle, not God in the flesh.

"Christ is the Word of God, therefore Scripture is not the Word of God" is an un-theological, modern sentiment—a sloppy play on words. There is no dichotomy between Christ who speaks and the Scriptures which are Christ's Word. On the contrary, there is the greatest harmony— Christ the Word is grasped by faith in words. We only find the true "Christ the Word" in physical words. "The term Logos ('Word') is used in the New Testament about two hundred times to indicate God's Word written, and seven times to indicate the Son of God—the Living Word of God."[14] And nowhere are they placed in opposition.

Christ is the concrete nexus and center of Scripture, not an irretrievable and undefinable axiom. "It is not enough to say that the [time-bound] human words point to a [timeless] Word beyond all human expression."[15] All Christomonistic attempts must be rejected. In them Christ can only find room to live in the individual's critical imagination: "The object of our faith is not any statement about Christ but the Lord

---

[10] G. Goldsworthy, *Gospel-Centered Hermeneutics*, 65.

[11] J. A. Zahm, *Evolution and Dogma*, 230.

[12] A. Richardson, *The Bible in the Age of Science*, 91.

[13] H. D. McDonald, *Theories of Revelation*, 1:265.

[14] L. S. Chafer, *Systematic Theology*, 1:72.

[15] Gordon R. Lewis in *Inerrancy,* ed. N. L. Geisler, 263.

Jesus Christ Himself."[16] He is no longer allowed to be the sin-bearer or to speak for Himself. An un-normed theology is inherently atheistic, normed only by reason's speculations. "The norm and standard for portraying [Christ] is the revealed Word and if one departs from this, he portrays not Christ, but his own dreams."[17] The methodological contrast is sharp: "the sixteenth century writers never understood . . . *doctrine* to be *revealed* in Christ, but rather in the text of Scripture."[18]

The bare statements of Scripture are not compelling for modern man, because in his heart he is Socinian: he expects the truth to agree with his thoughts. Moderns crave the academic freedom to criticize and make all history submit to their judgment. When Abraham Calov (d. 1686) attacked Socinianism in thousands of pages, he also condemned moderns. He "bitterly attacks this [Socinian] view as utterly un-Christian, and conducive to skepticism and even atheism."[19] Cherished methods and principles keep Christ at bay, while moderns poke fun at "dead orthodoxy" because of its "rationalism." If pre-moderns were simple-minded fundamentalists, moderns can only be atheists in comparison.

While "the Early Church's Christological model neither opposes nor separates Christ and Scripture," modern theology makes it an imperative.[20] For example: "The orthodox fathers did not provide a reason for why the biblical content was Christological."[21] The reason—not the rationalistic proof—is that Christ is God and we only know God through the seemingly weak words of Christ. It is truly rationalism to require decisive proof of unity and inner coherence, instead of submitting simply because God said. The unity of Scripture is to be assumed and believed, since divine truth, including the inner workings of the Trinity, is not open to man's investigation. The Spirit and the Son cannot

---

[16]World Conference at Lausanne (1927), quoted in: Theodore Engelder, "Holy Scripture or Christ?," 2 Parts, *Concordia Theological Monthly* 10:7–8 (July–Aug. 1939), Part 1, 496.

[17]Johann Conrad Dannhauer (d. 1666), quoted in: R. D. Preus, *Inspiration of Scripture*, 1.

[18]Richard A. Muller, *After Calvin: Studies in the Development of a Theological Tradition* (Oxford: Oxford University Press, 2003), 100.

[19]Fred Kramer, "The Inerrancy of Scripture: How it has been Understood, Attacked, and Defended," Faculty study paper (Copy in Concordia Theological Seminary Library, Fort Wayne, IN, 1964), 6.

[20]D. M. Farkasfalvy, *Inspiration and Interpretation*, 58.

[21]D. P. Scaer, "Theology of Robert Preus," 85.

contradict one another. They are together, with the Father, one God. Christ Himself says that the Spirit "will glorify me, for he will take what is mine and declare it to you" (Jn. 16:14).

Christ does not delegate speaking to men, because He wants us to be firm. One professor claims that "my grounding of the authority of Scripture in Christ is simply a Lutheran *fundamentum* [starting point]."[22] However, Christ is not a rational principle or internal standard in man. His authority is not vague and unknowable, but found in specific, revealed words. How does the authority of Christ, who fills all things, reach man? Is it in the simple words of Scripture or the spiritual enthusiasm of skilled scholars? The Lord directed His life not by abolishing, but by fulfilling written Scripture: "For the Son of Man goes as it is written of him" (Mk. 14:21).

Christ entered history to die for sins. We know this because of the testimony of the Spirit in Scripture, how all was done so that "what was spoken by the prophets might be fulfilled" (Mt. 2:23). Christ the Word actually spoke words, and He cannot deny Himself: "Do you think that I cannot appeal to my Father, and he will at once send me more than twelve legions of angels? But how then should the Scriptures be fulfilled, that it must be so?" (Mt. 26:53–54). His betrayal and death were authorized by Scripture—and that cannot be merely a human opinion to which Christ submitted. It is His own Word!

If the "Scriptures derive their authority from Christ, his resurrection and ascension, and the proclamation about him," any speech mentioning those words and concepts would have to be accepted, no matter how much they contradict Scripture.[23] Definite statements on human morality would be impossible to substantiate by this narrow gospel-authority. The modernist deconstruction of authority has radical implications for faith. Faith, where there is no divine word, is a "decision" or "crisis," which is entirely man-centered, and therefore atheistic, but faith according to Scripture is a miracle, not *in* the past, but *in* believers, so the Word is understood and comprehended. In this way the living Christ Himself is grasped and the sinner is made new before God. Christ's righteousness is granted in the Word, by which the Spirit works faith in the hearer. Faith is in the Word, and nothing outside or beyond it.

---

[22] J. Kloha, "Response to Montgomery Essay," 14.
[23] J. Kloha, "Response to Montgomery Essay," 14.

One can know of Christ derivatively from men, because the Gospel is propositional. The Word of Christ can be retold in many ways so that it remains true and effective. The Spirit can work through speech, writing, braille, sign language, and newly invented languages. The Gospel must go out into all the world: "And this gospel of the kingdom will be proclaimed throughout the whole world as a testimony to all nations, and then the end will come" (Mt. 24:14). Every sermon is to preach the same Gospel, but not in the same exact words. Yet, for a pure knowledge of Christ, Scripture is the only source we have been given. It is currently the only fountain of divine knowledge. Only Scripture is the norm for Christ. It is Christian to say, along with St. Bernard of Clairvaux, that Christ is the "*auctor* [author] of the scriptural text."[24] There is only one God, so the Spirit never acts against the Son of God.

"The Bible itself gives no key with which to distinguish the Word of God and Scripture, and along with that, between Scripture and Christ."[25] To assume a fissure in Scripture is to posit two different words of God. Man will be lord over both, since he is the one who authorizes with "divine" authority. The divine Bible evaporates before critical man. Those who attack the manger clothes of Christ will have nothing to proclaim except the six letters of the word "Christ." The doctrine of hell and the commandment against adultery do not sound very "Christological" to reason's enthroned ego-Christ, but since Christ spoke words on those topics, they are eminently Christological to the scriptural Christ.

Modernists say that fundamentalists have "little consideration of *how* authority functions in the economy of salvation."[26] This "new" approach is nothing but warmed-over 1970s Seminex[27] theology, which taught that "the inspiration of the written Word pertains to the effective power of the Scriptures to bring men and women to salvation through the Gospel."[28] Authority, though, is not functional, merely accomplishing

---

[24] D. M. Farkasfalvy, *Inspiration and Interpretation*, 179.

[25] G. Maier, *End of the Historical-Critical Method*, 16.

[26] P. H. Nafzger, *These Are Written*, 133.

[27] Concordia Seminary in Exile (later Christ Seminary–Seminex) was formed in 1974 after the walkout of the overwhelming majority of the faculty and students of LCMS Concordia Seminary, St. Louis.

[28] Faculty of Concordia Seminary, *Faithful to our Calling, Faithful To Our Lord* (1973), quoted in: P. A. Zimmerman, *A Seminary in Crisis*, 95.

things. If a parent believes in purely functional commands to children, it does not matter what is said or if it is obeyed. An authority that does not require any submission is imaginary.

An erring, wordless authority allows critical atheism. In the framework of economic-authority, scriptural doctrine is the ultimate legalism. Christ's words and teaching are given to graciously restrict us to Christ alone. Those who think without Scripture do what God incarnate did not dare to do! Christ bound Himself to letters and words.

To ask "why" the scriptural command was issued, or to get into the biblical writer's mind, is actually disobedience. Critics are like polite, disobedient children when they ask: "Why was this written?" The starting point of true interpretation is not "why" but what God has actually written. Authority resides in words, not some unknowable effect that may or may not occur even if the words are ignored. A father who allows a rebellious child to ignore his words is not using his authority, even if the child's behavior ends up being acceptable.

To make truth dependent on its salvific effect is backwards. Authority is not weighed and measured—it is obeyed or disobeyed. Parental authority, like scriptural authority, often does not have the desired result. Any authority of Scripture apart from its words is simply the freeway to unbridled criticism and unbelief. A purely functional truth based on Jesus' authority is not the scriptural Christ, but "the song of Jesus in our own hearts."[29]

"The living voice of the Gospel was for Luther the touchstone and measuring stick of all doctrinal judgments."[30] This is false, but has the ring of truth. The Gospel, as important as it is, has nothing to do with judgment. A norm both approves and condemns, but the Gospel is the application of righteousness. It does not accuse or condemn anything, but declares the forgiveness already won in Christ's death. A condemnatory gospel is an oxymoron. A negative, legalistic gospel is not connected to Christ's sacrificial death.

Life in Jesus is the heart of Scripture. To go beyond the words of God means leaving the truth and embracing a new legalism. The same spoken words inspired by the Spirit of Christ long ago remain just as

---

[29] T. Engelder, "Holy Scripture or Christ?," Part 2, 579.
[30] H. Gensichen, *We Condemn*, 144.

valid today: "I am astonished that you are so quickly deserting him who called you in the grace of Christ and are turning to a different gospel—not that there is another one, but there are some who trouble you and want to distort the gospel of Christ" (Gal. 1:6–7).

The real question is: "What defines the Gospel?" The modern hatred of all authority has caused distrust even of Scripture. The word "Gospel," like the letters which form the words "Christ," tell us nothing about what the true Evangel is. A timeless, eternal, ineffable gospel which has nothing to do with this world cannot be trusted. The Gospel is not only for God, it is for man to believe. Historical distancing is a facet of anti-authoritarian thinking. It is a power play to gain control of the Bible and its binding doctrines. The supposed human part of Scripture must be trumped up to overshadow God the Spirit.

If "the authority of the Scriptures do[es] *not* stand or fall with inspiration and inerrancy," their authority has nothing to do with reality or words.[31] If science has more objective standards of errancy and inerrancy, it is the higher truth. Christians, therefore, must be much more certain of the Gospel than the ever-changing theories of science. We do not have an errant Christ or an errant Gospel. A non-inspired truth is so limp it cannot even hold itself up as something to believe in the face of death. The Gospel itself, without a basis of authority, will only be a wet noodle which approves of every licentious activity.

To move beyond fundamentalism—beyond absolute authority—is to move past the true God. If the Scriptures "form in themselves the absolute touchstone of everything to be said and done in Christianity," no freedom is left for man to be god.[32] This fundamentalism is exactly the confession made by Lutherans subscribing to the Book of Concord: "Holy Scripture alone remains the only judge, rule, and guiding principle, according to which, as the only touchstone, all teachings should and must be recognized and judged, whether good or evil, correct or incorrect."[33]

When the Gospel is played against factual knowledge, it becomes otherworldly. Neo-orthodox or "dialectical theology can be understood to have an inherent tendency of dissolving [stable] theology into [spec-

---

[31] P. H. Nafzger, *These Are Written*, 134.

[32] J. Barr, *Beyond Fundamentalism*, 3.

[33] FC Ep Rule, 7; *The Book of Concord,* eds. R. Kolb and T. Wengert, 487.

ulative] Christology."[34] Without clear words of God, Christology is centered on man's sinful view of Christ, rather than Christ's sure words about Himself. Scripture is "the swaddling clothes and the manger in which Christ lies."[35] When the word-clothing of Christ is minimized or discarded, so is Christ. In a beautifully rendered statement of Ignatius of Antioch (d. 108), he describes "fleeing to the gospels, as the flesh of Christ." In essence, Ignatius means that "in order to understand the will of God, he fled to the gospels, which he believed no less than if Christ in the flesh had been speaking to him; and to the writings of the apostles, whom he esteemed as the presbytery of the whole Christian Church."[36] Luther says the same:

> The Holy Scripture is the Word of God, written and (as I might say) lettered and formed in letters, just as Christ is the eternal Word of God cloaked in human flesh. And just as Christ was embraced and handled by the world, so is the written Word of God too. It is a worm and no book, when compared with other books.[37]

It has become fashionable to belittle Scripture, assuming that Christ is thereby uplifted. "In sum, biblicism has become a blight on the American intellectual scene through . . . its self-defeating presuppositional circular-reasoning of biblical inerrancy and infallibility."[38] But the be-

---

[34] Uwe Carsten Scharf, *The Paradoxical Breakthrough of Revelation: Interpreting the Divine-Human Interplay in Paul Tillich's Work, 1913–1964* (New York: Walter de Gruyter, 1999), 87.

[35] Luther, quoted in: J. M. Reu, *Luther and the Scriptures*, 29.

[36] *Philadelphians* 5:1, quoted in: T. H. Horne, *An Introduction to the Critical Study and Knowledge of the Holy Scriptures*, 87. The standard translation sounds different only because of the modernist understanding of Scripture: "I have taken refuge in the gospel as the flesh of Jesus and in the apostles as the presbytery of the church. And we also love the prophets, because they anticipated the gospel in their preaching." An abstract "gospel" contrasts sharply with the definite apostolic and prophetic writings. "Flesh" is something that can be touched and grasped. Assuming such harmony between the "gospels" that they can be considered as one singular "gospel" is only possible if the Spirit is their author. *The Apostolic Fathers: Greek Texts and English Translations,* ed. Michael W. Holmes (Grand Rapids: Baker Books, 1999), 179.

[37] *Sprüche aus dem Alten Testament* (1541), quoted in: M. Thompson, *A Sure Ground on Which to Stand*, 105.

[38] John Bombaro, "Biblicism and the Imminent Death of American Evangelicalism," in *Built on the Foundation of the Apostles and the Prophets: Sola Scriptura in Context: The Second International Symposium on Lutheran Theology,* ed. Tapani Simojoki (Evangelical Lutheran Church of England, 2013), 163.

littling of the Scriptures *is* the belittling of Christ. Without Scripture enfleshing Christ, God Himself is put into a non-intrusive, man-made, Christological box.

The teaching of Christ is drawn from Scripture and leads the believer into Scripture as the purest source of truth. Scripture testifies of Christ. He used the Old Testament in all its apparent weakness to proclaim the Gospel on the way to Emmaus. Christ could have pointed to Himself and said: "I'm the true Word, so ignore Scripture."

Christ submitted to Scripture. There is little danger of taking the Scriptures too seriously, because they "point beyond themselves to Jesus Christ."[39] One can have faith in the scriptural Christ without the reading of Scripture (as with small children), but to defend Christ with conviction requires real authority. To say that inerrancy means faith in every fact of the Bible is to confuse God's unlimited authority to speak on any matter with the specific power of the Gospel that releases from sins: "It is beyond the purview of the gospel, let alone the Bible, that trusting Christ as the world's rightful king should entail equal faith in every single word in the Bible."[40] If Christ's authority is limited only to forgiveness, He is not truly God. Luther, however, makes a very pre-modern appeal to the fact of revelation, not by its content, but by confirmation in divine signs:

> You will find that these hardened blasphemers [who say that Scripture is only written by men] put themselves to shame and confusion with their own folly. They cannot even distinguish between a man who speaks for himself and one through whom God speaks. The words of the apostles were committed to them by God and confirmed and proved by great miracles, such as were never done for the doctrines of men.[41]

To appeal to a Christ outside of Scripture is to make Him a mental idol, and act as a pope who speaks for an absent Christ. One must know the words of Scripture to defend the teachings of Christ well.[42]

A personal, non-verbal authority is all too convenient. It puts no personal limits on man. Even before the New Testament was written,

---

[39] D. W. Lotz, "Luther and *Sola Scriptura*," 259.
[40] J. Bombaro, "Biblicism," in *Built on the Foundation,* ed. T. Simojoki, 166.
[41] *Avoiding the Doctrines of Men (1522)*, LW 35:152.
[42] S. W. Becker, *The Foolishness of God,* 164.

Christ and Scripture

the Church never went without the Old Testament Scriptures. Preaching and teaching require God's authority. It is "impossible to sever 'juridical' from 'personal' truth."[43] Christ is the truth and the life, but that life is revealed and defined in rationally understood words.

Christ is the norm of theology, but Christ is not a mute who needs us to speak for Him. He has given us a reliable Word, and the believer should hear Scripture as though the incarnate Christ is speaking to him. The scriptural Jesus is not stuck in a particular time, history, and context, like us, but has given us a sure and reliable Word on which to depend. The authority of the Bible "must be grounded in Jesus," but not the Jesus of the pious theologian's imagination.[44] The real Jesus "opened to [men] the Scriptures." Also, "he opened their minds to understand the Scriptures" (Lk. 24:32, 45).

Christ does not denounce or condemn error through the Gospel. The scriptural Gospel makes man a new creation before God by taking away sins. It does not condemn anything in man, but applies the righteousness earned by Christ's sacrifice personally. Grace is not a tool to determine right and wrong, truth and error. Forgiveness is no divining rod—it never holds back mercy or reproves sin. "Luther's Christological canon criticism" is a myth perpetuated by scholars who cannot appeal to Scripture.[45] If they can twist God's Word, Luther words are no challenge. If Scripture is a shadow of the eternal Word of God, but not the words of God, we are left with nothing at all. Christ becomes an unknowable, sub-human, impersonal idea.

People are so offended by Scripture, they are willing to pervert the Gospel. The true Gospel only grants freedom. What kind of gospel sets down limits, restrains thoughts, and says "no"? "For freedom Christ has set us free; stand firm therefore, and do not submit again to a yoke of slavery" (Gal. 5:1). The loss of Scripture alone is closely tied to the loss of justification in Christ. These "Christological scholars" are so wise, they cannot even define the Gospel, except as a sledgehammer to be used against Scripture. When the choice is to believe supposedly erring words of God or the fanciful words of human imagination, sinners will

---

[43] G. Maier, *End of the Historical-Critical Method*, 30.

[44] P. H. Nafzger, *These Are Written*, 145.

[45] Stephen J. Hultgren in *Rightly Handling the Word of Truth*, ed. C. E. Braaten, 36.

always pick the more convenient truth.

The real question of biblical statements is not "why," but "is it actually true, and can it be trusted?" Inerrancy is not a philosophical assumption, but a theological confession concerning the Bible's Author. "If God is the principal author of Scripture, and God, who is infinite Truth, can only assert what is true, then it follows deductively that the Word of God in Scripture can only contain truth."[46] Test God and His promises, and whether the Spirit can work through Scripture's words. Christ is real. The resurrected Lord speaks and hears—He is not a mute idol. "Jesus opened now his mouth, he who had previously opened the mouth of the prophets."[47] Divine power is not found in mentally manipulated ideas, but in Christ's very Word.

---

[46]S. W. Hahn, "For the Sake of our Salvation," 33.

[47]St. Bernard of Clairvaux, quoted in: D. M. Farkasfalvy, *Inspiration and Interpretation*, 214.

Chapter 28

Weak Analogies Prove Nothing

Pre-moderns often said the Bible is like Christ—in how it is treated, not its nature. Moderns use the axiom of the two natures of Christ to treat Scripture like the Pharisees treated Jesus' body. They test it and try to catch it in a contradiction. To claim the Bible actually has two natures is a direct outgrowth of modernist assumptions. The Christological analogy is completely unfounded scripturally, yet it seems to have found universal acceptance among academic theologians. It has been the downfall of much of the theological verbiage concerning the doctrine of Scripture.

The Christological analogy sounds very good, but it simply does not work. The Son of God took on human flesh in an eternal union. His human nature was sinless. The human writers were sinners. All their speech did unite eternally with God's Word. Writing Scripture was a one-time proposition—not a continual union of divine and human. The analogy of Christ to Scripture does not prove anything. At best it can be used to illustrate some aspects of the already solidly established scriptural doctrine.

While comparisons of the Bible to Christ's divine and human natures are not new, only since the rise of historical criticism have they been seen as determinative for establishing, or rather destroying, the nature of Scripture. "The theory of verbal inspiration contradicts the analogy of the revelation in Christ's person."[1] Revelation can only be in the irretrievable past to moderns, so the scientific fact of Christ is played against the scriptural Christ. "We cannot pass over the mystery of the *intermixing* of man's word with God's Word. The attempt to inquisitively unravel this intermixture and ultimately divide it," as historical criticism does, denies the mystery and cannot result in *any* divine doctrine.[2] The emphasis on the human aspect leads to assuming human diversity within Scripture, which means it cannot be God's words.

The false doctrine of the dual nature of Scripture is based on the philosophical incompatibility of human words and divine words. Human words cannot *be* divine, so Scripture is said to have two distinct, non-touching natures. However, if the eternal words are not the same as the human ones recorded for us, then the divine ones are inaccessible. What is left is the "humanity" of Scripture, which means in practice that it can be treated like any other writing. The actual humanity of the Scriptures is evident to all. It can be seen by the unbeliever that the Bible is written in human language. Very little needs to be said about it. Only emphasis on its divine authorship, by virtue of its inspiration, allows the Bible to be read as a single "letter" or literary work.

The two natures analogy is a false starting point, because it denies Scripture's divine origin. The acceptance of Christ, however, is the reception of His truth, which was breathed out in words. The only answer for believers is to be fools for Christ—to be considered "intolerant, narrow-minded, fanatics, [and] letter-worshippers."

> People are often more afraid of characterizations which diminish their honor and reputation than of the persecutor's sword. Very few are those who can accept the sacrifice of being considered stupid. But in today's world it is inevitable that every true Christian will be characterized as a fool, or at-least narrow-minded. . . . The world tolerates only those so-called Christians who walk in step with it, those who try to apply a social Christianity and attempt to be always up-to-date. The others who do not agree to adulterate their Faith it hates. But the world's hate is a criterion for us to know if we are true Christians. "If they have hated Me, they

---

[1] "Cambridge University Church Society: The Inspiration of Holy Scripture," *Cambridge Review* 5 (1884), 376.

[2] G. Maier, *End of the Historical-Critical Method*, 70.

shall hate you too."[3]

Christ's words have fared no better than His flesh.

> The New Testament testifies that Jesus experienced such things as fatigue, hunger, astonishment, grief, and extreme distress. Ultimately, he "humbled himself" to the point of accepting death, "even death on a cross." Yet virtually none of these empirical observations, which verify the full humanity of Jesus, force us to conclude that Christ divested himself of his divinity or surrendered his inherent impeccability. The weaknesses apparent on the surface of Jesus' historical life do not cancel or diminish his unseen perfection. At no point did he cease to be "the truth," the sinless and guileless Word of the Father. The same is true of the written Word of God. Despite its concrete expression in human language—even plain and sometimes imperfect language—it does not cease to be the divine discourse of God.[4]

The great Lutheran scholar Hermann Sasse, by starting with an improper Christological analogy, coordinated his theology to some of the assumptions of modern science. "[Sasse] thought that he was rendering a service by showing the church how and to what extent it ought to accommodate itself to some of the conclusions of the historical-critical methodology."[5] It was not the conclusions of modernism, but the very starting point—its definition of history and truth—that were alluring to him. Sasse spent much effort on the doctrine of Scripture. While he saw the significance of inspiration and inerrancy, he was not able to bring clarity to the doctrine of Scripture. With his substantial mental powers he could not bridge or synthesize the basic modernist–fundamentalist conflict over authority.

"Sasse's search for a third way [between fundamentalism and modernism] concentrated on the 'human side' " of Scripture. Unfortunately, his learned, though ultimately confused, writings on Scripture are still widely referenced. Modern Lutherans follow him in the search for the "holy grail" of a modern, intellectually fulfilling doctrine of inspiration,

---

[3] Alxander Kalomiros, *Against False Union: Humble Thoughts of an Orthodox Christian Concerning the Attempts for Union of the One Holy, Catholic, and Apostolic Church with the So-called Churches of the West*, trans. George Gabriel (Seattle: St. Nectarios Press, 1967; rev. 2nd ed., 1990), 51–52.

[4] S. W. Hahn, "For the Sake of our Salvation," 39.

[5] Eugene F. Klug, "Holy Scripture: The Inerrancy Question and Hermann Sasse," *Concordia Journal* 11:4 (1985), 127.

while his stronger suits, like church fellowship, are rarely mentioned.[6]

Christ comes to man; God, the Maker of reason, does not allow man to climb into heaven by his own mental efforts. It can only be intellectual works righteousness to not simply receive and believe the words which are Spirit and life. The modern cannot see the divine and human as one, so that "the divine and the human, the eternal and the temporal, exist side-by-side."[7] Sloppy applications of the Christological–scriptural analogy rule modern theology.

Sasse, though, accurately saw the big picture:
> We are in the situation in which the Church found itself when the Arian Controversy arose. . . . No one liked the word *homoousios* [of one substance]. Even Athanasius used it very sparingly. But sometimes the Church has to coin a term to denote a certain truth. The word was meant to safeguard the full divinity of the Son of God. . . . *Inerrancy* has become the bone of contention in our time between Lutherans. Years may pass until the patient work of theologians will lead to a fuller and better understanding, just as in the case of the *homoousion* in the 4th century. The term *inerrantia* [inerrancy] cannot and should not be given up—the meaning is quite clear, the absence of real error in the Bible.[8]

Unfortunately, his theological starting point was un-Lutheran: he wanted to undo all the history of the Church and start with an ill-suited analogy. If the Reformers missed the mark, their Confessions are wrong and 2,000 years of Christian witness to Scripture's inspiration must be reversed. This modern thinking is dangerously arrogant and completely un-orthodox. The fear of too close an identification of God and man lead men to rationally try to balance the unbalanceable. In the words of another Lutheran scholar, the Bible as the written, absolute Word of God is sinful to moderns—it "becomes a docetic book."[9]

It is tragic that for all of Sasse's historical insight, he was a child of

---

[6]Kurt Marquart in *Hermann Sasse: A Man for Our Times?*, ed. J. R. Stephenson, 185. Sasse strangely thought that, "modernism is the natural, legitimate child of Fundamentalism. . . . We are not supposed to win the young people entrusted to our care to a certain world view (*Weltanschauung*), but to turn them to their Savior . . . ." As it turns out the scriptural Christ is inseparable from the biblical worldview. H. Sasse, *Scripture and the Church*, 164.

[7]G. Ebeling, *Word and Faith*, 38.

[8]H. Sasse, Letter to J. A. O. Preus (1970).

[9]T. H. Rehwaldt, "The Other Understanding of the Inspiration Texts," 360.

his time and unable to come to grips with the teaching of inspiration because of his modernist convictions: "The philosophical foundations for our scholarly, theological work, was that our approach to the world that we examine and penetrate in different ways will always be decided by the present, admittedly always changing, state of natural science." It is scientific assumptions and natural philosophy which made adopting Luther's full confession of Scripture tragically problematic for Sasse. He worked with a secular definition of knowledge: "Luther would never have approved what Pieper still thinks of the natural knowledge of God." The scientific view of history, which allowed for his great insights, also changed Sasse's basic approach to God's Word. He wrote to an orthodox LCMS theologian: "Another difference in our theological thinking is to be found in the different understanding of history." He claimed that Walther sadly did not know modern historical scholarship, which is critical by nature.[10]

Under the name of "historical science," theology, especially for Germans, has become a critical endeavor, rather than bare facts which are either true or false. The case of Sasse shows that even when one's theological conclusions are well-received, historical-critical assumptions will change the basis of theological work. Theology can never be a completely scholarly endeavor, because "scholarly" implies certain critical and anti-confessional tendencies. To be scholarly is to be scientific, and even atheistic to some degree.

While moderns are uneasy with revelation's divine authorship, "orthodoxy has always accented the divinity of the Scriptures."[11] There is no need to say much about their humanness—it is plainly evident. Any desire to improve orthodoxy, that is, what Christ has revealed, is doomed. Thinking great thoughts for God in order to rationally develop doctrine is a very generous offer by modern man, but it is a denial of the sufficient truth revealed in Scripture, which we have before our eyes.

So while Sasse criticized Aquinas as a rationalist, he himself partook of atheistic historical distancing in the name of "scholarly" knowledge. This issue of Scripture and its inspiration, however, was not a problem for Aquinas, but was included under the general category of "prophetic

---

[10] Tom G. A. Hardt in *Hermann Sasse: A Man for Our Times?*, ed. J. R. Stephenson, 156, 161, 162.

[11] H. J. Eggold, "A Man's View of the Word of God," 1.

speech."[12] Sasse tried to find a way to divine truth through the human author, while Aquinas was content to say that "God is the author of Scripture."[13]

The "humanity" of Scripture is the avenue through which critics render judgment on it. "We dare not speak too much of Scripture's lowly form, lest we quickly become lords of it."[14] Without tangible, human words from Him, Christ is simply an idea, a philosophical principle that allows man to determine the Gospel. This Gospel will always turn into law, since reason is vitriolic in its hatred of Jesus and His truth. In not believing the scriptural incarnation, reason "invents a God after its own fancy. Reason agrees that God's Word is to be honored, but arrogantly sets itself up as a judge deciding what is, [and] what is not, God's Word."[15]

Those who do not allow criticism of the Bible's words are said to hold a "docetic" view, implying a denial of its "human nature." The docetic charge, however, more accurately applies to Zwingli (d. 1531) and moderns who hold that the divine Word is incompatible with human words. Zwingli warned against an "irreverent deification of the created thing," that is, the external Word, since "one does not have the 'inner Word'' just because he has Scripture or its propositional content.[16] To external man, Christ only comes externally in words. Here is the true root of the divine–human separation of Scriptures: a cleaving of the Spirit from the external Word.

Analogies are not doctrine, nor can they establish anything certain.

---

[12] "It is requisite to prophecy that the intention of the mind be raised to the perception of Divine things: wherefore it is written (Ezek. 2:1): 'Son of man, stand upon thy feet, and I will speak to thee.' This raising of the intention is brought about by the motion of the Holy Ghost, wherefore the text goes on to say: 'And the Spirit entered into me . . . and He set me upon my feet.' . . . Accordingly inspiration is requisite for prophecy, as regards the raising of the mind, according to Job 32:8, 'The inspiration of the Almighty giveth understanding': while revelation is necessary, as regards the very perception of Divine things, whereby prophecy is completed; by its means the veil of darkness and ignorance is removed, according to Job 12:22, 'He discovereth great things out of darkness'." *The Summa Theologica of St. Thomas Aquinas* (2nd rev. ed., 1920; http://www.newadvent.org/summa/3171.htm), 2.2.171.1.

[13] Quoted in: M. Thompson, *A Sure Ground on Which to Stand*, 32.

[14] G. Maier, *Biblical Hermeneutics*, 185.

[15] Luther, quoted in: B. A. Gerrish, *Grace and Reason*, 14.

[16] Quoted in: A. A. Zaun, "A Study of the Idea of the Verbal Inspiration," 194–195.

Doctrine is from God, and no man can accept the revelation of God without spiritual enlightenment. Moderns are too proud to begin with Scripture, so they begin with their own thoughts: "Assuming that God's words are of one piece with, and congruent with, one another, it is, therefore, best to understand the nature and attributes of the sacred Scriptures by means of the incarnation, i.e., Christologically."[17] Where the doctrine of God does not rule, man legislates against Christ. This is simply modernism, right in line with atheistic neo-orthodoxy: "The Church must develop its doctrine of the Scriptures on the same lines as the doctrine of the two natures."[18] Christ, like the doctrine of Scripture, is not a man-made idea. The early creeds on the God-man Christ precisely follow the words of Scripture, or else they cannot be confessed as divine truth.

The result of building a personal, theological system is that Scripture and truth are called idols and legalisms. "To regard the bible as intrinsically authoritative easily degenerates into idolatry."[19] In contrast, true Christianity confesses that the Bible is "not a burdensome law but the very Magna Carta of Christian liberty," because it gives true knowledge of the resurrected Christ.[20]

Certainty is a basic characteristic of faith. Luther could say without any qualms: "I am certain that the Word of God is with me and not with them, for I have the Scriptures on my side and they have only their own doctrine."[21] This knowledge did not originate from man, nor did his certainty depend on secular knowledge. Those who are uncomfortable with the divine certainty of Scripture appropriate the scientific model of accuracy, which is never complete or certain: "A better way out [of historical uncertainty] is reliabilism [not infallibilism, or absolute certainty] which posits degrees of reasonableness, instead of the false dilemma of either certainty or uncertainty."[22] This man-centered approach to knowledge and certainty is deadly to faith. As

---

[17] J. W. Voelz, *What Does This Mean?*, 233.

[18] E. Brunner, *Revelation and Reason*, 276.

[19] Warren Quanbeck, quoted in: R. A. Bohlmann, *Principles of Biblical Interpretation in the Lutheran Confessions*, 130.

[20] J. Gresham Machen, quoted in: S. J. Nichols and E. T. Brandt, *Ancient Word, Changing Worlds*, 57.

[21] *Defense and Explanation of All the Articles* (1521), LW 32:9.

[22] M. R. Noland, "Walther and the Revival of Confessional Lutheranism," 216.

Philip Melanchthon describes,
> [reason] does not see or touch what is proper to the Gospel, that is, the forgiveness of sins to be given without recompense, for the sake of the Son of God. This notion has not sprung from human minds, indeed, it is far beyond the range of human reason, but the Son of God, who is in the bosom of the Father, has made it manifest.[23]

While reason has always been the troubling enemy of theology, it was never accepted as the method of doing theology until the modern era.

As applied to inspiration, "mechanical" is not a descriptive theological word. It has rightly been called the "stumbling-block" of modern theology. It gives the connotation of "impassive machines," as if the human authors had "written without, and contrary to, their will, without consciousness and unwillingly."[24] Scripture's divine and human authorship is not a matter of balance. Doctrine is not man's to balance. It is to be accepted and used. No serious Christian theologian has denied that Scripture was written by men in human language. To do so, one would have to deny Scripture itself, since it mentions human authors extensively. For example: "I, Paul, write this greeting with my own hand" (1 Cor. 16:21).

"The whole period following the general repudiation of inerrancy and the introduction of higher criticism has been marked by an attack on the 'mechanical' theory of inspiration." "Mechanical dictation," that is divine authority, is an affront to man's divine status. The writers directed by the Spirit are viewed by moderns "as mere *machines,* possessed of no power of [free] choice in the selection and use of words they used."[25] God's direct intervention with man is most offensive to moderns, because it implies that something is lacking in his natural state. The very idea of revelation implies that non-revealed knowledge and reason are insufficient.

The main import of inspiration is that God is also the author of the words written by men. There is no awkward coordination. God and man did not alternate writing Scripture verses. The result of inspiration is a single, divine, readable authority for teaching Christ. The scriptural

---

[23] *Loci* (1555), quoted in: *The Devil's Whore: Reason and Philosophy in the Lutheran Tradition,* ed. Jennifer Hockenbery Dragseth (Philadelphia: Fortress Press, 2011), 75.

[24] T. Engelder, "Verbal Inspiration: A Stumbling Block," Part 12, 483, 486.

[25] H. D. McDonald, *Theories of Revelation,* 2:218.

writings are patently different in style, language, character, and mood. The evidence says different men wrote them. It is an act of faith to confess their unity—not a result of intense, validating scrutiny. There are plenty of human words about Christ; the one thing that sets Scripture apart is its divine authorship.

The Christian position on inspiration is summarized by Justin Martyr: "even when people speak in Scripture in answer to God in Scripture, it is the Divine Word [Christ] who speaks."[26] However, in 1670 Baruch Spinoza, who "laid the groundwork for the 18th-century Enlightenment and modern biblical criticism," asserted: the writers were "only inspired when speaking directly the words of God."[27] Following Spinoza, a conservative Lutheran can now rationalize:

> that 'the Word of the Lord came' . . . is a clue that, as there must be a distinction between the prophetic word and the direct Word of God to the prophet, so there must be a similar distinction between the direct Word of God and the apostolic word. . . . If the prophetic word is always the Word of God in the same sense as the direct speaking of God . . . then in every conversation between God and a prophet God would be speaking with Himself.[28]

Where supposed levels of inspiration are said to differ, man must step in as the true divine revelation and decide the highest truth. The pre-modern St. Ambrose (d. 397), however, saw no difficulty in saying, "so Moses opened his mouth and uttered what the Lord spoke within him."[29] Inspiration as something less than God speaking is meaningless as a theological term. Luther's assertions, on the other hand, are based on the fact that "the Holy Spirit is no fool or drunkard, who would speak one iota, much less a word, in vain." "Wherever in Scripture you find God speaking about God, as if there were two persons, you may boldly assume that three Persons of the Godhead are there indicated."[30]

Denial of divine inspiration leaves only human writings, or at most an indeterminable, synergistic mix. In the two natures analogy there is some blend of human and divine cooperation that must be carefully

---

[26] Quoted in: W. L. Craig "Men Moved by the Holy Spirit Spoke from God," 158.
[27] *Wikipedia, The Free Encyclopedia,* https://en.wikipedia.org/wiki/Baruch_Spinoza; W. L. Craig "Men Moved by the Holy Spirit Spoke from God," 167.
[28] D. P. Scaer, *Apostolic Scriptures,* 51.
[29] Quoted in: S. Rose, *Genesis, Creation, and Early Man,* 127.
[30] *On the Last Words of David* (1543), LW 15:280.

divided by the critic. But the human aspect of Scripture is what allows truth to come to man. The very human details of Scripture, such as external circumstances, prior written sources, extensive research, and autobiographical references, are fully compatible with divine inspiration. Luther had no problem with this hypothesis: "I believe that Adam wrote for several generations, and after him Noah and the rest, to describe what happened to them, then Moses."[31] The pre-history of possible scriptural sources did not affect his confidence in God's Word, because it was not based on his critical examination of the materials, but on God's declarations in Scripture. The Spirit can use sordid history and the fallibility of man for His holy will.

The genuine analogy for inspiration is the conversion of the sinner. Man is purely passive in receiving salvation, but also willing and cooperating. He is not forced or mechanically coerced—yet the Spirit does all the work and deserves all the credit.

> It is correctly said that in conversion God, through the drawing of the Holy Ghost, makes out of stubborn and unwilling men willing ones, and that after such conversion in the daily exercise of repentance the regenerate will of man is not idle, but also cooperates in all the works of the Holy Ghost, which He performs through us. . . . man's will in his conversion is *pure passive* [purely passive], that is, that it does nothing whatever, . . . when God's Spirit, through the Word heard or the use of the holy Sacraments, lays hold upon man's will, and works the new birth and conversion. For when the Holy Ghost has wrought and accomplished this, and man's will has been changed and renewed by His divine power and working alone, then the new will of man is an instrument and organ of God the Holy Ghost, so that he not only accepts grace, but also cooperates with the Holy Ghost in the works which follow.[32]

In the same way the writers (the instruments and organs of the Spirit) were commanded to write, but did so willingly. They "cooperated" in the process, using their natural vocabulary, but in such a way that every word is completely from the Spirit. Modern man, who buckles at inspiration, is likely to also have a problem with conversion and faith. The same Spirit who inspired the Scriptures also renews sinners through

---

[31] He even offers scriptural proof: Num. 21:14; Josh. 10:13. *Table Talk* (1540), LW 54:373.
[32] FC Ep II:17–18.

the Word and sacraments. He moves, inspires, and carries them to good works.

The divine–human analogy is the starting point for a large number of modern theologians today. One LCMS professor wrote, tellingly: "I have learned from my teachers (by writing, if not firsthand) Sasse, [Martin] Franzmann, and [James] Voelz to understand the Scriptures as both divine and human. This has profound implications for how we hear them."[33] This sounds very spiritual, but it exchanges Christ's words for the words of sinners. In 1886 Walther said that critically dividing the mythical divine and human sides of Scripture was satanic. He compared it to Zwingli's *Alleosis,* which Luther ridiculed as positing two Christs. Zwingli attributed statements about Christ to either the divine or the human nature, never the entire person, in effect making two distinct persons. In the same way, Walther accurately saw that starting with a disharmony between the supposed "divine–human natures" in Scripture leaves one with two worthless scriptures.

> Beware, beware, I say, of this "divine–human" Scripture. It is the devil's mask. For eventually it constructs such a Bible after which I would not wish to call myself a Bible Christian. Henceforth the Bible is nothing more than any other good book which I must read with constant and diligent examination lest I be counseled in error. For if I believe that the Bible also contains errors, then it is no longer a touchstone for me, but needs a touchstone itself. In short, it is unspeakable what the devil tries with the "divine–human" Scripture."[34]

The result is that the "human side" is open to critical judgment, while the "divine side" is beyond description. There has been an enormous change in doctrine within the LCMS from Walther's day. Divinely authoritative doctrine has been replaced by vague, ill-fitting analogies. *Sola Scriptura* is not even a relic of tradition to revere—it is dismissed as sub-Lutheran and rationalistic. What is the benefit for modern man? When Scripture is less than divine, a critical approach (man himself) is needed to procure the divine. He authorizes himself to speculate and theorize, instead of believing and confessing.

---

[33] J. Kloha, "Theological and Hermeneutical Reflections," in *Listening to the Word of God: Exegetical Approaches,* ed. A. Behrens, 181.

[34] Robert D. Preus, "Walther and the Scriptures," *Concordia Theological Monthly* 32:11 (1961), 674.

We find God in His Word. Christ's truth is not spoken or understood without the Holy Spirit. His Spirit not only moved the writers, but animates every faithful proclaimer of Christ.

> It is the devil himself whatsoever is extolled as Spirit without the Word and Sacraments. For God wished to appear even to Moses through the burning bush and spoken Word; and no prophet neither Elijah nor Elisha, received the Spirit without the Ten Commandments or spoken Word. Neither was John the Baptist conceived without the preceding word of Gabriel, nor did he leap in his mother's womb without the voice of Mary. And Peter says, 2 Pet. 1:21: The prophecy came not by the will of man; but holy men of God spake as they were moved by the Holy Ghost. Without the outward Word, however, they were not holy, much less would the Holy Ghost have moved them to speak when they still were unholy or profane; for they were holy, says he, since the Holy Ghost spake through them.

Luther calls this separating of God the Spirit from His external, objective Word "the strength of all heresy," but now, proved by a man-made analogy, it has become the starting point when moderns deal with God's Word.[35]

---

[35] *The Smalcald Articles* (SA) III.III:11–13.

Chapter 29

Historical Criticism in the LCMS: Examples

A most un-theological view is that the fight for the Word of God is over. Many assume that because battles were bitterly fought in the past, they do not rage on today. One LCMS district president recently said that "historical criticism [in the LCMS] was eradicated in the 1970's."[1] But God's Word is always under attack by Satan's birds (Lk. 8:12). Without the divine Word, Christ is not trusted or known.

The LCMS in principle upholds Scripture and its inspiration. In practice, the Scriptures are not directly used as the only authority. A consequence is pacifism. We tolerate false doctrine and what is unscriptural without a way to reject it by divine right.

The reason the LCMS is no longer able to confess clearly is found in her teachers. Many professors of her seminaries and colleges are guilty of despising God's Word, but as leading scientific experts they seem to be above criticism. Criticizing a credentialed scholar or professor is the ultimate academic sin. We desperately need polemics—born out of love for the truth. To tolerate truth and error equally is to despise both. "Whoever is not with me is against me, and whoever does not gather with me scatters" (Mt. 11:30).

Neutrality and objectivity towards the teaching of Christ is not

---

[1] A verbal presentation of Iowa East District President Brian Saunders at the Nebraska Lutheran Confessional Series conference, Lincoln, NE (Dec. 10, 2015).

Christian. "The modern shift to the human [scriptural] authors helped liberate exegesis from polemical pathos."[2] Every individual is either under Satan's control or has been brought by the Word into Christ's marvelous light. Whatever endangers or harms this salvation and the pure doctrine which brings Christ is cursed by God. "We destroy arguments and every lofty opinion raised against the knowledge of God, and take every thought captive to obey Christ" (2 Cor. 10:5). It is the duty of those who have the truth to defend it, out of love for Christ—even if it means fashioning a whip to chastise those sitting in the temple of a seminary. "To condemn is the highest service of love to the erring themselves."[3]

We Missouri Synod Lutherans are kings of paying lip service to inspiration, but it really does not influence our thinking or practice. This is seen in the attempts to uphold the word "inspiration," while at the same time asserting it is unhelpful and meaningless. "Assuming that we have an oracle-like, perfect text of the Bible is [ironically] convenient" to moderns, because it does not allow critical appropriation. One recent attempt is to say that the traditional manner in which we account for the inspiration and authority of Scripture is "rationalistic" and the same as in Islam, therefore it must be wrong. "We have an accounting that risks not being specifically Christian in that it is not clearly rooted in the work of Christ or the Holy Spirit."[4] Truth is made and authorized by man, not revealed, in modern theology. If a theological argument is not unique and impressive to reason, it carries no Cartesian weight.

Within the LCMS the inspiration of Scripture by the Spirit is said to be a 19th century invention. "I disagree with [Theodore] Engelder's [1865-1949] argument that the authority [of Scripture] is based primarily on an act of inspiration that is embodied in the 'autographs' (which, of course, do not exist)."[5] This disagreement is not with Engelder, but with Luther and the whole of Lutheran theology. The lack of autographs only irritates the critic, who does not have them for rational examination. For

---

[2] Karlfried Froehlich in *Studies in Lutheran Hermeneutics,* ed. J. Reumann, 140.
[3] H. Gensichen, *We Condemn,* 179.
[4] J. Kloha, "Theological and Hermeneutical Reflections," in *Listening to the Word of God: Exegetical Approaches,* ed. A. Behrens, 184. The original essay states it much stronger: "We have an accounting that is not specifically Christian, nor rooted in either Christ or His Holy Spirit . . . ." J. Kloha, "Text and Authority," 8.
[5] J. Kloha, "Response to Montgomery Essay," 13.

him authority is not based on an act of God, so it must necessarily come by human verification. So the scholar who feels compelled to examine God's Word as a neutral critic, must rail against inspiration, calling it a "super-naturalistic, docetic understanding of the Scriptures."[6] Another seminary professor (Kloha's declared mentor) says: "one cannot become 'Docetist' on this matter: the Scriptures still are products of human authors writing from a particular perspective at a particular point in time. . . . the divine and human authorships of the books of sacred Scripture always present themselves in creative tension."[7] This "tension" means that God cannot ever fully mix with men, because "thus saith scholarly man."

Even outright denials of any meaningful accounting of inspiration can be accepted in the LCMS, if the exegesis is reasonably traditional: "The work of the Spirit in giving us the Scriptures is not, strictly speaking, a direct or immediate work."[8] One may question whether this pronouncement was spoken under a better inspiration than Scripture's. No proof is offered except that it seems reasonable to the "devil's whore," reason. What role does the Spirit actually have in the Christian's life if He was not even present directly in the writing of Scripture? "But if the Spirit of Him who raised Jesus from the dead dwells in you, He who raised Christ from the dead will also give life to your mortal bodies through His Spirit who dwells in you" (Rom. 8:11). The direct role of the Spirit in conversion and faith is simply collateral damage in the all-out attack on Scripture.

Authority becomes the only legalism, because it restricts atheistic criticism. Scripture as judge over man's words and thoughts is explicitly called law: "the legal analogy is telling; Scripture is being viewed not as a living, active Word of God, but a static, divine, authoritative, propositional legal text."[9] The modern cannot see anything as timelessly true and at the same time human. "Static," unchanging words are said to be the opposite of the Word of God. What is spiritual and living would not dare restrict the highest thoughts of man—it is falsely assumed. But Luther's explanation of the Third Commandment depends on God's

---

[6] J. Kloha, "Theological and Hermeneutical Reflections," in *Listening to the Word of God: Exegetical Approaches,* ed. A. Behrens, 192.

[7] J. W. Voelz, *What Does This Mean?*, 242.

[8] D. P. Scaer, *Apostolic Scriptures*, 65.

[9] J. Kloha, "Text and Authority," 12.

Word being something that is handled and reissued by preachers, while remaining fully God's Word: "We should fear and love God so that we do not despise preaching and His Word." To despise propositionally correct preaching is to deny God and profane His commandment.

When propositions are not made, sermons become story time and the conviction of divine authority is lost. A proposition is "a statement that affirms or denies something."[10] A non-propositional Scripture says nothing that can direct or reprove the thoughts and lives of men. Non-propositional doctrine and sermons can only tickle the imagination. Authority is exercised in words and commands, not in creative thoughts. The true authority of Scripture judges propositionally to protect the Gospel and Christ. Christians certainly can know right and wrong behaviors. They are to judge between correct and false teachings. To judge is a legal duty, but "shall not the Judge of all the earth do what is just?" (Gen. 18:25) What kind of god only approves and never condemns? One which Marcion (d. 160) formed by excising the Old Testament and most of the New Testament. Those who alter Scripture's words fashion a more desirable god in their own image.

Even supposed conservative stalwarts would like to redefine Scripture: "Inspiration is not an autonomous act of the Sovereign God, but an extension of the incarnation."[11] "The Spirit who inspires the Scriptures does not immediately come from the Trinity, but is present in the community having broken into history through Jesus of Nazareth." In this view God is present only historically in Christ, leaving the Scriptures open to scientific analysis. The community of sinners (man), then, takes the place of God's sure Word and knowledge. "If God encased himself in history," "any direct intervention of the Spirit in the life of the church, including the inspiration [of the Scriptures], gives the appearance of a *deus ex machina* [something convenient that is unexplainable or unnatural], and militates against a Christological definition of doctrine."[12] Despite his protests, man does not have authority to define doctrine from his own personal conception of Christ. Speculation is not divine revelation. Christ has defined Himself for us in Scripture. Having ready-made truth

---

[10] *American Heritage Dictionary of the English Language* (5th ed.; http://www.thefreedictionary.com/propositionally).

[11] David P. Scaer, *An Introduction to the Method and Practice of Lutheran Theology* (Fort Wayne, IN: Concordia Theological Seminary Press, 1990), 27.

[12] D. P. Scaer, "Biblical Inspiration in Trinitarian Perspective," 158.

in the Bible, directly from Christ's Spirit, is too easy for modernists, who want to form theology according to their own idolatrous image of what seems most "Christological" or "Trinitarian." God's actual involvement with man, even in defining theology, is ruled out as unreasonable.

The "Christological view of inspiration" seems promising. Pre-moderns, however, did not include "the historical origins of the biblical documents in their doctrine of the Bible as the Word of God."[13] How does history, critically determined by sovereign man, improve on the Spirit's words? Were the disciples closer to Christ than His own Spirit? It is in actuality sinful historical-critical methodology driving this satanic urge to evolve the doctrine of Scripture. It is readily admitted: "*Redaktionsgeschichte* [redaction criticism], puts the emphasis on the writer, somewhat in the sense as apostolic authority," which takes the place of divine inspiration.[14] The disciples of the radical Bultmann invented this method of interpretation that limits the references of the various biblical writings to the Christian communities to which they were addressed. The biblical writers are not seen as inspired, but as erring and questionably motivated "redactors" (editors). Then the Scriptures tell us about the human writer's revisions of Christ and his audience's motivations, but not about Christ directly. This indirect "Christology" must be teased out by the professional exegete, who has academic license to widely spout unprovable speculation.

The modernist does not find inspiration meaningful because direct revelation is incomprehensible to him. So "the doctrine of inspiration is not necessarily *wrong*. But it is not the most helpful way of thinking about the Bible." This author then flatly denies any meaningful use or understanding of the Bible: "Instead of thinking of inspiration as a past event or a guarantee of ontological [real or actually existing] perfection, it is better understood as the work of the Holy Spirit in and through the Scriptures." What is the substitute for divine words? God might speak through man's word, "where and when God chooses to speak through it." " 'Deputized discourse' occurs when one person speaks *in the name of* another person."[15] The doctrinal position today in the LCMS is not all that different from the Seminex theology of the 1970's. At the St.

---

[13] D. P. Scaer, "Theology of Robert Preus," 85.

[14] D. P. Scaer, *Apostolic Scriptures*, 62.

[15] P. H. Nafzger, *These Are Written*, 17, 41, 68.

Louis seminary, at that time, "no one denied verbal inspiration." It was tolerable, "if you avoid literalism."[16] Like today, the liberal St. Louis professors "all agreed that the Holy Scriptures are authoritative," but they could not agree on the basis of its authority. The argument was made that it is "the Christo-centric character that tells him that the Scriptures are God's word."[17] Scripture, though, is either based on the breath of the Spirit or men of dust.

Sinful man's word is not a substitute for the divine word. God's designated messengers, both apostles and prophets, are to be ignored if they do not speak divine words. A more Lutheran teacher assails modern Lutherans: "It is cursed unbelief and odious flesh which will not permit us to see or know that God speaks to us in the Scriptures and that it is God's Word, but tells us it is the word merely of Isaiah, or Paul, or some other mere man who has not created heaven and earth."[18] Is a "minister of the Word" even possible without divine words spoken in human language?

The true, justifying Gospel is Christ's own Word. It explains and delivers the benefit of His historical actions. To rely on the Word of God is to rely on Christ Himself: "The word is near you, in your mouth and in your heart, that is, the word of faith that we proclaim" (Rom. 10:8). The Word of Christ, which can be reformulated in preaching, is the only link to Christ. It is not far away, though it is often refused in willful disobedience. Spiritually dead men prefer an absent, non-speaking God. This is their idea of freedom: to treat the Holy Scriptures as their own personal Mad Lib, filling in the divine blanks with their own insights.

In truth, this doctrine of Scripture, while not the center of theology, does affect all doing of theology. A wrong understanding of inspiration will unsettle all doctrine eventually. So, in supreme arrogance, all of Church history, including the Nicene Creed, is cast aside: "Neither liberal criticism nor the doctrine of inspiration is sufficiently rooted

---

[16] A Concordia Seminary, St. Louis professor (1970–71), quoted in: P. A. Zimmerman, *A Seminary in Crisis*, 173.

[17] P. A. Zimmerman, *A Seminary in Crisis*, 174.

[18] Luther commenting on Is. 55:11. *Auslegung vierler schöner Sprüche* [Exposition of Many Beautiful Verses], *Dr. Martin Luthers Sämmtliche Schriften* (St. Louis: Concordia Publishing House, [1880–1910]), 9:1800; trans. Robert D. Preus in *Inerrancy and the Church*, ed. J. D. Hannah, 117.

in the Gospel of Christ crucified."[19] The Bible is said to be less than sufficient for giving all knowledge of Christ, but we have books and essays by sinners which purport to help ground the authority of Christ in something better than the Holy Spirit. To attempt to do theology better than Scripture is the satanic arrogance of one who tries to out-theologize Christ Himself. All attempts to creatively remake doctrines are dead from the start. Doctrine is to be received, or it cannot be divine. It is not mental clay to be molded by the whims of sinners. The urge "to account for the Bible on naturalistic principles" is part of the modernist plot to remove God from this world.[20] All teachings, yes, even the doctrine of inspiration, are related to Christ. To attack the least of these is to attack Him who spoke them.

Peter Nafzger's book *The Cruciform Scriptures* puts forth a theology not just outside the bounds of Lutheranism, but outside Christianity. It flatly denies any relevance for the historic doctrines of inspiration and inerrancy: "the authority of the Scriptures do[es] *not* stand or fall with inspiration and inerrancy." Not only does Nafzger deny *sola Scriptura*, he does so in a most flagrant way: "The meaning of a text does not reside solely within the text itself." If the Bible is not actually God's words, God remains "completely free," per the existential theology of the Reformed Karl Barth. The human words might "become the Word of God," but truth itself is not something to be grasped or communicated.[21] Why do great scholars burden us with their blathering speculations, if the truth cannot be defined in words? Perhaps their words are a more suitable vehicle than the Spirit's words. Crucifying the text of Christ is not preaching Christ crucified to real sinners.[22]

Righteousness in Christ is received by the faith which the Spirit produces. That faith is dynamically created by static, meaningful words. The same Spirit who brings Christ inspired the words in which Christ's salvation is authentically defined for all time. Not coincidentally, one of the strongest confessional statements on Christian freedom is linked to the Spirit's speaking: "Likewise, the article concerning Christian liberty also is here at stake, which the Holy Ghost through the mouth of the

---

[19] P. H. Nafzger, *These Are Written*, 160.
[20] H. D. McDonald, *Theories of Revelation*, 2:131.
[21] P. H. Nafzger, *These Are Written*, 134, 147, 44.
[22] They are "crucifying [the text] on the cross of their own opinion." Luther, *The Bondage of the Will* (1525), LW 33:181.

holy apostle so earnestly charged His Church to preserve . . . ."[23] This view seems to disregard the human instrument, but in doing so allows pure, absolute truth to be praised. The unity of the biblical words allows the one, scriptural Christ to be known, but the modern Lutheran, who starts with his own sight, assumes that all "attempts at harmonization inevitably blur [the gospels'] theological and literary distinctiveness."[24] The diverse perspectives of men are considered more important than God's voice.

Even the newest standard translation of the Lutheran Confessions seems to misunderstand Lutheran theology because of critical assumptions. When Melanchthon refers to 1 Cor. 11:29 to prove the condemnation which comes upon unworthy recipients of the Lord's Supper, the Bible verse is introduced with "Christ says." This is a mighty theological point: Christ Himself determines who takes His Supper. This certain knowledge is only possible through the inspiration of the Spirit, who spoke Christ's words. But this critical edition, quite ignorantly, has a footnote with a correction: "Actually, Paul."[25] If they are merely Paul's words, who would use this one verse to exclude the unworthy and those confessing a different doctrine of Christ? The condemnatory eating of the unworthy, part of the basis for the practice of closed Communion, is based directly on this one verse. Christ did speak through Paul, as the Confessions say, therefore this teaching still applies today and divinely guides pastors in admitting to Christ's Sacrament. To deny the teaching expressed in 1 Cor. 11:29, on the other hand, is to go against not just St. Paul, but Christ Himself.

The Confessions assume a single theology. They speak of "the pure doctrine of the divine Word," while in modern thought, the divine–human tension leads to multiple human theologies.[26] Man's words in the Confessions are then the brighter lamp to lead one into the dark minefield of God's Word: "The Confessions as Hermeneutical Guides Themselves." The Lutheran Confessions do have value for unclear readers. Yet if the clear Scriptures do not verify the Confessions, they cannot

---

[23] FC SD X:10.

[24] Peter J. Scaer, "The Gospel of Luke and the Christology of Martyrdom," unpublished paper presented at the Fort Wayne Exegetical Symposium (2003; http://static1.1.sqspcdn.com/static/f/38692/333371/1273663586093/The+Gospel+of+Luke), 1.

[25] AP XI:5; *The Book of Concord,* eds. R. Kolb and T. Wengert, 186.

[26] FC SD Rule, 3.

be accepted without reservation: "What can be said of the earliest *regulae fidei* [rules of faith] and the creeds cannot be said with reference to the 16th-century Lutheran Confessions from the Augustana to the Formula of Concord." The distance from the Christ (not Scripture) is claimed to make more recent confessions less certain.[27] This inability to commit to any confession absolutely is a symptom of reason's supposed sovereignty, which does not want to be bound. But a partial confession is no confession at all. There is no half-confession of Christ, who is Lord of all.

The Confessions of Lutheranism understand themselves as doing exactly what the early creeds did: putting God's Word in human words to refute troubling false doctrine:

> And since of old the true Christian doctrine, in a pure, sound sense, was collected from God's Word into brief articles or chapters against the corruption of heretics, we confess, in the second place, the three Ecumenical Creeds, namely, the Apostles', the Nicene, and the Athanasian, as glorious confessions of the faith, brief, devout, and founded upon God's Word, in which all the heresies which at that time had arisen in the Christian Church are clearly and unanswerably refuted.[28]

The 16th century Lutheran confessional documents claim to witness to the same eternal, revealed truth as that of the earliest creeds. The age of a human formulation of the truth is irrelevant, if it is the eternal truth of God's Word. Scripture is propositionally used to destroy false knowledge, so that it is God's own judgment against error. This should be a basic Christian activity of every generation, since no human words are eternally sufficient.

What is Christian is only decided by judging all human writings by divine Scripture. Scripture is not necessary for faith, but it is required for accurately defining what doctrinal content is to be believed. It is "the only true standard by which all teachers and doctrines are to be

---

[27]"The basic treatment of John 6 is also possibly an issue with the Confessions." Perhaps this is because upholding faith, a direct *action* of the Spirit, over the visible *act* of the Supper is inherently problematic for modern Lutherans. It is irrational to them that Christ talks about eating without also teaching about the Supper. In Christ's Scripture, however, faith in the Word of forgiveness has priority over the sacramental eating, which can only lead to condemnation when done without faith (1 Cor. 11:29). J. W. Voelz, *What Does This Mean?*, 358, 349.

[28]FC SD Rule, 4.

judged."²⁹ The Confessions did not start with trying to balance God and man within the pages of Scripture. This was unthinkable in pre-modern Christianity: "In efforts to defend the divinity of Scripture, is it possible we have implicitly downplayed their equivalent humanity."³⁰ There is no separate divinity in Scripture apart from the humanity of words. We only truly deny their "humanity" when we cease to read and use them. The greatest danger is that we judge them like all other human writings.

> Dr. Luther himself . . . has expressly drawn this distinction namely, that the Word of God alone should be and remain the only standard and rule of doctrine, to which the writings of no man should be regarded as equal, but to which everything should be subjected.³¹

In the fact of inspiration neither God nor man are canceled out or opposed. The Spirit who was breathed into man to make him can certainly move Him to willingly write. Man's word is fully God's Word, without limitation, in the "the manifest Scripture of the Holy Ghost."³²

As a result of weakness in the LCMS on the doctrine of Scripture, Barthian language and concepts are in vogue: "Revelation is . . . God's free decision in eternity to be *our* God." This unknowable and wordless revelation is not a revelation in any meaningful sense—it certainly cannot be preached to a crowd. "God speaks through Scripture," but the Bible is not here identified with the divine Word because "the Scriptures become the Word of God only where and when it pleases God."³³

The celebrated Reformed scholar Barth carried the axiom that "the finite (man) is not capable of the infinite (God)" to the ultimate extreme. Nafzger describes Barth's position (in propositional form!): "God cannot be defined in static terms," meaning doctrine is not fixed in any confession or human words.³⁴ God is only an activity or series of events to Barth, so no language can speak about Him. This definition of a changing God, whom doctrine cannot contain, has nothing to do with

---

²⁹FC SD Rule, 3.

³⁰Erik Rottmann, "Sermon of the Western Missouri Pastoral Conference: Preached at St. Paul High School, Concordia, MO" (unpublished; Oct. 17, 2006), 2.

³¹FC SD Rule, 9.

³²AC Preface, 9.

³³P. H. Nafzger, *These Are Written*, 47, 44.

³⁴P. H. Nafzger, *These Are Written*, 47.

God's own words about Himself. Moderns of this ilk follow in the steps of Barth, "who turned his back on the metaphysics of classical theology rendering almost unintelligible the conceptual idiom of . . . [the] creeds of the Church."[35] God has been so distanced from humanity that He is only a figment of the imagination of man—a ridiculous reflection of the changing culture.

If language is incompatible with the Spirit, all doctrinal statements are meaningless—incapable of holding God's infinitude. "The Sacred Scriptures, for all their divine inspiration and all their august authority, were inevitably conditioned by the languages in which the Holy Ghost inspired the words of the Divine Revelation and by the circumstances that evoked them."[36] While the attack is against Scripture, Christ is the one who suffers. The statement "God became man" is a proposition that can be stated correctly in different phrases and languages. If Christ cannot be defined in static terms, Christ is no longer the incarnate mediator. In fact, another savior would be required to bridge the historical and philosophical gap between the historical Christ and modern man. Without Scripture, Christ's own Word, there is no contact point with the physical Christ.

The main difference between Barth and Rudolf Bultmann is only that Barth used Christian terminology to create a man-made system immaculately tailored in traditional verbiage. Bultmann came to the same basic conclusion, but did so with integrity and clarity. Because of radios and other technology, he said: "the only honest way of reciting the creeds is to strip the mythological framework from the truth they enshrine."[37] This is the Enlightenment project carried to its conclusion: complete and utter darkness. If there is no actual Word of God, there can be no true speaking about God. The Barthian position is that "the entire activity of the triune God is subsumed under the rubric of revelation."[38] Buried under an avalanche of words, the Gospel for Barth

---

[35] M. Rose, "Karl Barth's Failure," 44.

[36] Arthur Carl Piepkorn, "The Significance of the Lutheran Symbols for Today: 1954 Faculty Lecture Series," in *The Sacred Scriptures and the Lutheran Confessions: Selected Writings of Arthur Carl Piepkorn,* vol. 2, ed. Philip J. Secker (Mansfield, CT: CEC Press, 2007), 2:85.

[37] Rudolf Bultmann, *New Testament and Mythology and Other Basic Writings,* trans. Schubert Miles Ogden (Philadelphia: Fortress Press, 1984), 4.

[38] C. E. Braaten, *History and Hermeneutics,* 13.

is without content and unhistorical, defined only by personal action. Barth tellingly wrote in a letter: "How frighteningly indifferent I have become about the purely historical questions."[39] It is not scholarly to repent, but can there be any action other than to reject such nonsense?

"Christ is risen from the dead" is a proposition about God. To deny this proposition the ability to speak about earthly facts and history is to put oneself outside the doctrinal bounds of Christianity (1 Cor. 15:12–19). Karl Barth, the theological patron of Peter Nafzger, put the resurrection in "realm of the supra-temporal and supra-historical," so "it need not be thought of as following chronologically Christ's death" or a part of "recoverable history."[40] It avoids critical analysis and rational denial, but also any hope of factual truth. Modern theologians quickly leave scriptural Christianity in their denial of Scripture: "The original event of Christ's resurrection belongs to the historical past and has no ultimate significance for me. . . . In this case, the basic factuality, is declared to be a matter of indifference, in the name of the historically significant."[41] Whoever holds this must turn from their methodological atheism, no matter how many "Christological" words flow from their pen. The loss of propositional revelation entails the sure loss of Christ.[42] One LCMS seminary professor calls Barth's idolatry an "impressive accomplishment."[43] The Christ of Scripture does not agree, and He is not afraid to critique the geniuses of modern theology. "Woe to you, scribes and Pharisees, hypocrites! For you travel across sea and land to make a single proselyte, and when he becomes a proselyte, you make him twice as much a child of hell as yourselves" (Mt. 23:15). Woe to you, hermeneutical philosophers and exegetes, hypocrites, who lead away from the scriptural Christ!

Scripture is almost universally denied by modern Lutherans, though in a very subtle way: "Scripture as a static storehouse of timeless propo-

---

[39] Quoted in: C. E. Braaten, *History and Hermeneutics*, 22.

[40] H. D. McDonald, *Theories of Revelation*, 2:91.

[41] H. Thielicke, *Prolegomena*, 110–111.

[42] Pieper rightly said that inspiration and atonement were linked, so that when the Bible falls, so does the heart of Christianity. He does not state the reverse, that inspiration guarantees a doctrine of Christ. Revelation is claimed by many religions, but only one speaks of Christ.

[43] Foreword by Joel P. Okamoto, P. H. Nafzger, *These Are Written*, ix.

sitional truths denies its historical character."[44] On the other hand, they suppose that real truth is found in science, whose propositions are either true or false. Deriding the propositional character of Scripture (in dry, propositional fashion) is endemic among moderns. The static property of Scripture is set in opposition to its (only potential) sacramental instrumentality: "Scripture is being viewed not as a living, active Word of God, but a static, divine, authoritative, propositional legal text."[45] It is wrong that "divine" and "authoritative" are played against "living" and "active." What Jesus said to those who denied the propositional doctrine of resurrected life applies here: "He is not God of the dead, but of the living. You are quite wrong" (Mk. 12:27).

Those who have renounced the "thesis that divine revelation contains truths are left without any rational basis for theology."[46] If God only reveals Himself non-propositionally, in a personal encounter, "our knowledge of God cannot be written down in a series of propositions, which can then be setup as 'the truth about God,' whether in the Scriptures," creeds, books of dogmatic theology, or sermons.[47] In this "personal" revelation nothing is actually communicated, therefore nothing is binding on man. No statement about Christ or morality has unqualified authority. As a result of this scientific atheism the LCMS is attracted to every faddish program and scent of new methodology. "The LCMS says nothing, imitates everything."[48]

The attempts to "correct" Walther, Pieper, and the entire doctrinal heritage of the LCMS are quite appealing to rationalists marinated in critical methodology since birth. "Many theologians may find out too late that a blow directed on the face Franz Pieper landed on the face of Christ."[49] The invitation to write a better dogmatics than Pieper remains unanswered by his critics. On the most pressing theological issue of the last few centuries, he was absolutely correct.

Moderns study a "problem," which means they are not confident

---

[44] L. Jordahl, "The Theology of Franz Pieper," 131.

[45] J. Kloha, "Text and Authority," 12.

[46] H. D. McDonald, *Theories of Revelation*, 2:97.

[47] A. Richardson, *The Bible in the Age of Science*, 90.

[48] Kurt Marquart in *Hermann Sasse: A Man for Our Times?*, ed. J. R. Stephenson, 192.

[49] John R. Stephenson, "'Inerrancy'—The *homousias* of Our Time," *Logia* 3:4 (1993), 5.

or comfortable with what Scripture plainly says. To state anything authoritatively without allowing any exception is deemed legalistic. However, God did speak. Scholars, steeped in doubt-inducing methods, are actually sheep. Their authority is nothing compared to God's very Word. Despite their critical methods, most are critical in exactly the same uncritical way. The same Seminex language of the Gospel as theological norm is heard today: "If the Scripture was thoroughly Christological . . . then the gospel was the standard in judging the Scripture." Why would Christ need to judge the Spirit's words, which are truly His? An undefined Gospel as the norm of theology is truly a rational circle, allowing no certain knowledge above man. Direct, revealed knowledge is forbidden by those who do theology "from below," from man's point of view.[50] The same root error of denying the existence of God's Word is cropping up in the LCMS. The results are more palatable at first glance, but the conviction is the same: "It is impossible to regard [Paul's letters] as oracles in a literalistic way."[51]

Historical criticism has been taken up uncritically. To say that "inspiration is proof of an event's historical character" is called "ahistorical."[52] Without an authority to submit to, it is an easy and burden-free task to theologize. It might require great learning to be a scholar, but to do theology only takes a little imagination. Themes and clever motifs allow one to overcome historical distance and the denial of all revelation. The Bible then talks about theology, but legitimizes nothing.

The result is hobbyhorse theology. Theologians ride the same cookie-cutter theme to death, wearing out the text. Everything is then read in terms of a predetermined thesis—the text does not set the agenda. The Word of God is not directly applied, but verses are read into preconceived themes. Every eating must be about the Lord's Supper and every drop of water must be about baptismal regeneration, regardless of what the text actually says. When God does not think like us, we bend the words to our authority. We humble God in the most egregious ways, and think we are doing Him a favor. In scholarly interpretation truth is in man, not the all-authoritative Lord. So where do theological themes originate, and what measures them, if not the words of Scripture? Tradition, liturgical

---

[50] D. P. Scaer, "Theology of Robert Preus," 88, 85.

[51] A Concordia Seminary, St. Louis professor (1970–71), quoted in: P. A. Zimmerman, *A Seminary in Crisis*, 173–74.

[52] D. P. Scaer, "Theology of Robert Preus," 84.

niceties, cutting-edge programs, and novel speculation rule—in short, anything but the eternally valid Word of God.

One Missouri Synod pastor holds the Bible's "*sola Scriptura* standing within the Church catholic," but insists that,

> A faithful and obedient tradition of interpretation is one that submits to Scripture's in-built hermeneutical principles presented to us by none other than Christ Himself, such that they are safeguarded and substantiated by multiple extra-biblical sources of corroborating authority (for example, the liturgy, creeds, and consensual patristic commentary on Scripture and the much neglected oral tradition known as "canon of truth" or the "rule of faith").[53]

The claim that Scripture is so weak that it needs supporting authority renders all scriptural authority moot. Scripture alone is what sets Lutherans apart—not just the claim, but the actual confession of what Scripture says. The "Church catholic" was, and is, wrong where it departs from *sola Scriptura*. The Church does not have priority or authority over her Head and Lord.

When the Gospel is said to critique and judge, it is no longer the free forgiveness based on Christ's death. The Gospel is played against Scripture in the modern LCMS:

> The Scripture is to be believed on account of Christ, its essential content and interpretive determiner. The other doctrine of biblicism holds that Scripture is trustworthy because of philosophical commitments that supposedly prove its divine origin by means of inspiration. Our heritage, our responsibility is to the former, not the latter.[54]

Is hearing and trusting Christ also a "philosophical commitment" to be avoided? Reason judges based on the observable results. A clear Scripture should be clear to all sinners, but its interpreters do not all agree, which forms hard, empirical "truth" that the Scriptures are dark and need support from man.

What takes precedence over Scripture?

> But the Scriptures understood in this [literal and fundamentalistic] way would mean that modern understandings of cosmic origins, genetics, psychology, the physical sciences, medicine, history, etc., are overruled and deemed irrelevant

---

[53] J. Bombaro, "Biblicism," in *Built on the Foundation,* ed. T. Simojoki, 155.

[54] J. Bombaro, "Biblicism," in *Built on the Foundation,* ed. T. Simojoki, 174.

> to faith by the Bible's purported understanding. . . . Such a view essentially forecloses on all intelligent discussion on issues which are, or seem to be, treated in the Bible.

Secular knowledge, which knows nothing of forgiveness, is the inerrant authority for such moderns. "Significantly, it is also [LCMS Seminary Professor] David Scaer who sets the course for corrective Lutheran action: . . . repenting of betraying its first love—Jesus Christ—for an idolatrous affair with the Bible."[55] It is only the modernist deification of man that permits any separation between Scripture and Christ. "For Luther *sola Scriptura* was in perfect harmony with *solus Christus.*"[56]

The desire to achieve intellectual respectability before men is deadly. One of the most alluring modern idols is avoiding the charge of biblicism, which is a form of fundamentalism to the modern. To be unthinking is far worse than to be unbelieving. Luther evidently does not have a say today in defining Lutheranism: "When the Holy Spirit speaks about insignificant and despised matters, He simultaneously enfolds most precious gems of the greatest virtues."[57]

Modernism has impressive words, but nothing certain, solid, or definite. Without an outward norm, every attack is potentially deadly for a church. Lutherans become reactionary: we do not want to appear Reformed, so we will ape the equally wrong Roman Catholics, or we do not want to look Roman, so we parody every practice of those who have no written confession. One must be careful that in his righteous criticism, he is not reacting against true orthodoxy. Truth is not something that can be rearranged or developed to suit our tastes. The following subtitle is utterly un-Lutheran and un-Christian: "Toward a Cruciform Theology of Scripture."[58] We either have a definite theology and confess it with divine conviction, or we should be silent. God is not an academic plaything. His Word is meant to be confessed by God's people. But it is implicitly denied in the scientific neutrality of the ivory tower.

One's scriptural method changes one's attitude towards God's Word and God Himself. There is no way to practice historical criticism without it affecting the doctrine of Scripture or the Gospel itself. This statement

---

[55] J. Bombaro, "Biblicism," in *Built on the Foundation*, ed. T. Simojoki, 165, 168.
[56] E. F. Klug, "Word and Scripture in Luther Studies," 32.
[57] Commentary on Gen. 30:2. *Lectures on Genesis*, LW 5:33.
[58] P. H. Nafzger, *These Are Written*.

by a "conservative professor" is replete with modernist thinking on Scripture: "The Gospel becomes the theological standard for judging doctrine." For another example, take this massive historical and theological blunder: "For Luther the Christological norm superseded the biblical one."[59] Unbelief is an inappropriate method to apply to promises of the faith. "So-called modern criticism . . . is unable to offer anything concrete, true, and real, because it despises the only source of truth, the divine Word."[60]

There is no divine blood shed for forgiveness, unless Jesus is involved in history, the same as any other person. Christianity is based on historical facts, but we can only know for sure what Christ tells us in Scripture. Where there is no certainty or comfort in the face of death and damnable sins, there is no Gospel—no matter how much Jesus-verbiage or talk about authority is vocalized. Repentance, without which there is no forgiveness, requires that we are first in opposition to God's will and truth. But in the dynamic redefinition of truth, no one is wrong—everyone is theologically right, no matter what propositions are made, as long as the cherry of "Christ" is laid on top. A consequence of modernism is that the substance of Christianity, that is, " 'to repent' has become 'utterly unintelligible'."[61] The only solution is to turn from being our own authorities and submit to Christ's scriptural authority. "Christ holds to the usage of Scripture and places before us an example how we ought not to speak, hold, or mention what is not founded in the Scriptures."[62]

---

[59] David P. Scaer, *Introduction to Method and Practice of Lutheran Theology*, 22.
[60] F. Bettex, *The Bible the Word of God*, 273.
[61] F. Bettex, *The Bible the Word of God*, 45.
[62] Sermon on Mt. 23:34–39. Martin Luther, *Complete Sermons of Martin Luther*, 7 vols., eds. John Nicholas Lenker and Eugene F. A. Klug (Grand Rapids: Baker Books, 2000; vol. 1–4 published as *Sermons of Martin Luther: The Church Postils*, 8 vols. in 4 vols., 1995; vols. 5–7 published as *Sermons of Martin Luther: The House Postils*, 3 vols. 1996), 1.1:227.

Chapter 30

Conservative Atheistic Approaches

Modern theology refuses to interact meaningfully with the hard sciences, society's presumed truth, because it capitulates completely to secular facts. Most theology, especially at the academic level, accepts the status quo of the kingship of scientific knowledge, but when it refuses to bridge theology with the perceived factual truths of science, it precludes the possibility of absolute truth. This explains the vitriol against apologetics by moderns: they have nothing to defend against secular knowledge. "Thus Barth (and [Dietrich] Bonhoeffer) considered apologetics misguided, because it transgressed the boundaries separating the empirical and religious realms."[1] To isolate scriptural study as a harmless academic subject is to deny implicitly its universality and absoluteness. "However much authority one cared to give to Scripture, tradition, or inspiration, they could not claim to be the sovereign or ultimate standards of truth, simply because it was possible to assess their authority according to reason."[2] The authority of reason replaces the Holy Spirit in scientific approaches to Scripture.

In this dead end of modern theology the interpretation of Scripture is isolated from dogmatics, preaching, and pastoral care by scientific

---

[1] Richard Weikart, "The Troubling Truth about Bonhoeffer's Theology," *Christian Research Journal* 35:6 (2012); republished: http://www.equip.org/PDF/JAF5356.pdf, 4.

[2] F. C. Beiser, *The Sovereignty of Reason*, 323.

methodology. Each theological sub-discipline becomes disjointed, with no divine truth to unite all theological thinking. Modern exegesis has become a predominately creative exercise, not a stabilizer of divine teaching. "Historical-critical methodology is accepted today by virtually all exegetes, regardless of their confessional background."[3] This universal, rational works righteousness requires implicit faith in its methods and principles. This "neutral method" is defended with religious fervor.

In silencing the Bible, "the exegetical department of the academy [is] the new magisterium of the contemporary church."[4] The main question for redaction critics is: "Why was this written?" To get behind the text and find the author's motives is acceptable for human writings (even though the answer is never without the taint of sin). But what of God's writings? To go behind the words (beyond what the text actually says), to a non-verbalized intention, is actually to psychoanalyze God Himself.

When a parent gives an authoritative command, the child may not ask "I wonder why he said that?" He is bound by God's command to honor his father and mother, so he may not even question their words. The questioning of authority is simply disobedience and sin. Human reason, however gifted, can never replace the Spirit who revealed the words of divine truth. Truth has its origin in the God of truth—not men. The search for naturalistic, human explanations for Scripture is simply a technique to bypass direct authority and put man on the same level as God. Truth is not to be explained away but adored and worshiped.

So whether a biblical writing was written to catechize, for worship, or for a certain human audience, the truth is much fuller. Scripture, God's book, was written for all, not just its original receivers. It does not address abstract, man-made themes, such as the sacramental, but establishes divine truth, contrary to this assertion: "Luther's failure to find a eucharistic allusion in John 6, the discourse on the Bread from Heaven with its requirement that believers eat His flesh and drink His blood, may seemingly provide Lutherans with a minimalist sacramental hermeneutic."[5] One may have all sacraments, outwardly, and not have Christ and the forgiveness of sins. Without facts, all rational connections

---

[3] John Breck, "Exegesis and Interpretation," 77.
[4] Ralph Bohlmann in *Studies in Lutheran Hermeneutics,* ed. J. Reumann, 204–205.
[5] David P. Scaer, "Reformed Exegesis and Lutheran Sacraments: Worlds in Conflict," *CTQ* 64:1 (2000), 14.

are only mental notions. A preconceived hermeneutic is a man-made idol used to obscure the clear Scriptures and submit them to man's will. The entire doctrine of Christ is established by the inspired Scriptures, and thus by God Himself. Scripture is not a microwave dinner serving what is most convenient for making lazy arguments today. It is God's will made known for all mankind.

The Scriptures have as much relevance for present readers as they did for their original receivers. It is not a dead, human document but the living voice of the Spirit, who spoke static truths. "When the biblical events are treated in isolation from one another, that is, not as a post-Easter reflection of the apostles in the life of the church that was born in baptism and was nourished by the Lord's Supper, a non-sacramental reading of the biblical texts is inevitable."[6] Such lofty themes are attractive to sinners who do not want to wade into the nitty-gritty basis of theological facts, but themes prove nothing. Finding undefined sacraments under every scriptural rock is not confessing the pure truth of Christ. Christians facing sin and death need solid meat, not artistic and thematic fluff.

In the thematic approach doctrine becomes assumed. This results in a lot of Christ-talk, using fancy, "spiritual" words like "Christology," "cruciform," and "Christocentric," but it cannot offer the living Christ—the Lord who was born of the virgin and offered Himself for the sins of the world. Jesus is not a hypothesis—He actually died and takes away sins with the Word of His power. Even if done in the name of Christ, critical methodologies have nothing to do with the real Christ. The biblical Christ gave His life for our peace and healing (Is. 53:5). He did not give Scripture as a plaything for scholars, but to enable sure knowledge of His salvation. "Scriptures for the exegete are fenceless prairies where he may roam, and so he is the envy of those whose goals are predetermined by tradition and official boundaries."[7] But the faithful "biblical interpreter is just the opposite of the 'neutral scientist'," who treats God's Word like an interesting chew toy.[8] He has God's words, which are not a light to be hidden.

Historical criticism naturally leads to multiple theologies. Each theme

---

[6] D. P. Scaer, "Reformed Exegesis," 18.
[7] D. P. Scaer, "Reformed Exegesis," 5.
[8] G. Maier, *End of the Historical-Critical Method*, 90.

becomes its own theology, offering its own salvation and Christ. This critical assumption, which denies the unitary authorship of Scripture, allows one to prove any position on the Supper—because nothing is provable if Scripture has contradictory threads of thought:

> Contemporary critical scholars, in distinguishing between the event and later theological interpretations of the event, attempt to identify the steps from the event, which for them is often unrecoverable, and the final form in the Gospels. . . . Such methods are not without value in that the earliest church reflections on the Lord's Supper are seen to resemble closely what later became the classical Reformed view of a symbolical meal. Texts in their final form, as we have them in the Bible, were encrusted with views now associated with Lutherans and Catholics. Because the Gospels preserve both earlier and later reflections on the Last Supper, Lutherans and Reformed justified their accommodation as biblical with each other on the Lord's Supper in the Formula of Agreement.[9]

God's Word is not a stable base for faith or knowledge in this view: Scripture is merely a human document with a multiplicity of differing theological strands, which the trained scholar must tease out. In fact, it is so unclear that being certain of what the text means is the only sin. This kind of Lutheranism would never have allowed Luther to reject the rationalist argument that Christ cannot be in the Supper. Instead, he rejected church fellowship with the deniers of the Supper based on one word of Scripture. Luther sang of the Devil: "One little word can fell him." He did not say "one hermeneutic" or "abstract theme" or "unprovable allusion," but that one word of Christ (such as "is") is sufficient for victory. Luther believed words made the divine accessible and that Scripture is the basis for all Christian knowledge.

That is not to say all instances or conclusions of historical criticism are heretical, but that the basic approach is anti-Christian. Any method that puts man and God on the same level is doomed to failure. "Man does not live by bread alone, but man lives by every word that comes from the mouth of the Lord" (Dt. 8:3). To ask "why" of a word of Scripture is to bite the hand that feeds all humanity. It is unbelief not to believe what God says. All questioning of Scripture and treating it as something that we can look critically down on is *sin*, a denial of the

---

[9]D. P. Scaer, "Reformed Exegesis," 19.

Lord Christ. Faith does not require knowledge of all of Scripture, but the denial its words tends to the denial of their Author. A little leaven leavens the whole lump (Gal. 5:9). A little unbelief in Christ's will is a dangerous precedent.

If no external authority is admitted, only the internal is left, but "an authority which has its source in ourselves is no authority."[10] Those Christians who deny inerrancy should be asked whether the infallible criterion of truth lies within themselves, if it is not found in Scripture. The drive to prove God's standard of inerrancy *by* man's standard of inerrancy is backwards and rebellious by nature. Truth in modernist thinking must be verified by man and no sacred exception is allowed.

Within churches extreme havoc is caused by the presumption that man is the critical arbiter of truth. It is called "biblicistic" when "the interpreter tries to jump back across two millennia which separate us from the writers of Holy Writ."[11] God's Word is said to be only for the trained, judicious scholar, not the common Christian. However, when God is believed to be the author, two millennia is a small feat for the eternal Son of God. Christ does it through the Holy Spirit—not in old, worn-out words, but eternal ones.

When one is free to judge Scripture and find gnostic, submerged treasures, no human norm can hold back the dam of criticism. The account of the woman at the well is a "well-crafted narrative," according to reason.[12] But the simplicity of Mark is a stumbling block to moderns, who are in love with inventing high-level, rationalistic themes. Luther did not struggle with its apparent weakness—he cites it without an asterisk in the Small Catechism (Mk. 16:16). The differences in the gospels make "no difference to the faith of believers, since in all of them all things are declared by the one sovereign Spirit."[13]

Scripture is not a gadget to be looked at objectively. It is academically respectable to meticulously analyze and dissect Scripture like a corpse, but that is a denial of Christ's ability to speak. The words of the Bible

---

[10] H. D. McDonald, *Theories of Revelation*, 2:304.

[11] Lowell Green, "Toward a new Lutheran Dogmatics," *CTQ* 50:2 (1986), 111.

[12] Peter J. Scaer, "Jesus and the Woman at the Well: Where Mission Meets Worship," *CTQ* 67:1 (2003), 4.

[13] *The Muratorian Canon* (2nd–3rd cent.), quoted in: G. Maier, *Biblical Hermeneutics*, 139.

bring Christ to lost sinners. Even if unbelieving reason cannot prove its internal harmony, it must submit to the authority of God. "The four gospel canon . . . posited the assumption that basic harmony reigns among the various canonical presentations about Jesus. It did not do so obviously by historical or exegetical demonstration," but by confession of the one Spirit who spoke.[14]

In some cases Scripture's words do have a pre-history, but any such excavation behind the words is not a theological activity. To base an interpretation on an assumed priority or historical ordering of gospels is extra-biblical and uncertain—a denial of *sola Scriptura*. At the appearance of contradiction, Luther "sought some expedient that might remove the difficulty and frequently in doing so ventured to propose daring hypotheses."[15] To go behind the bare words is to put oneself above God. Only in the fact of the Spirit's inspiration can there be reliable knowledge of God. Being a 1st century Jew did not lead to identification with Christ's understanding, or else none would have shouted "crucify him." In essence, biblical criticism is exegesis "undertaken without regard to its dogmatic consequences."[16] For a church that claims to have the pure doctrine, this approach is problematic. A replacement for the finality of "God says" can only be dubious and put a question mark behind all doctrine.

The result is that "the Gospels are not the life of Christ as much as they are the life of the early church."[17] "With the rise of form criticism and redaction criticism in modern times it has been realized more clearly than formerly that even the three Synoptics have their own tendencies in theology and spirituality. Nevertheless the 'synoptic fundamentalism' is still with us."[18] The modern focus on the human writers is at the expense of the Scripture's divine unity and authorship. "*Redaktionsgeschichte* [redaction criticism] says a lot about the author's intentions and how he accomplished them in the written gospels," to the detriment of their unity. On the other hand, direct inspiration is said to "circumvent" the incarnation, as if God can interact with His creation only at this one

---

[14] D. M. Farkasfalvy, *Inspiration and Interpretation*, 106.
[15] J. M. Reu, *Luther and the Scriptures*, 49.
[16] Alister E. McGrath in *The Science of Theology*, ed. P. Avis, 219.
[17] B. Ramm, *Protestant Biblical Interpretation*, 67.
[18] Arland J. Hultgren in *Studies in Lutheran Hermeneutics*, ed. J. Reumann, 147.

historical point.[19] This is warmed-over deism matched with scientific naturalism. The Spirit's act in inspiring Scripture is the only sure path *to* the truth of the incarnation. As a historical event critical man has no access to Jesus, as liberal theology has demonstrated. When the unity of Scripture is lost, coherent theology is impossible. Man's rational efforts are not needed to refine and translate Christ's Word. "While the blood of the martyrs may be the seed of the Church, as Tertullian affirmed, the sweat of the exegete can only be the seed of heresy."[20]

Narrative

Fiction is a modern category of literature which is neither true nor false. "Truth" is irrelevant to the idea of narrative. Stories preform no ruling or judging function. Fictional approaches, generally labeled with the word "narrative," seek to move beyond historical-critical methodology. Narrative theology does not repudiate the older methods, though, but builds on them. It is actually more radical, because it considers true, factual knowledge as not only uncertain, but irrelevant.

Historical criticism aims to find factual knowledge scientifically. Narrative theologians see the need for factual knowledge as a weakness, since historical criticism is essentially negative. Its view of truth is even more man-centered than the older criticism. "Narrative theologians did give primacy to story over doctrines." The genre of narrative avoids any intersection with factual history, thereby avoiding criticism of the facts of Christianity. "Thus the Bible can be true without being historically accurate."[21] With defenders of Scripture like this, what are the godless to critique?

Though aligned with "post-modernism," the narrative approach is a complete rationalizing and humanizing of theology. It is the search for coherence, not truth. Narrative theologians want to have it all: to avoid fighting, but possess the similitude of truth. Nothing is rejected, not even the most radical of historical-critical investigations. In its "revelation weariness," it wants to dispense with facts, history, and truth—to tell

---

[19] D. P. Scaer, *Apostolic Scriptures*, 64, 68.
[20] John Breck, "Exegesis and Interpretation," 75.
[21] S. J. Grenz and R. E. Olson, *20th-Century Theology*, 273, 285.

good (possibly untrue) stories.[22] "Propositional" is a modern code word for "doctrinal," signifying the burden of divine doctrine upon enlightened men.

One LCMS professor states in his narrative study: "In passing, I may note my own conviction that the story related in Matthew's Gospel is, in fact, historical." That is precipitated because his method, narrative criticism, treats everything as fiction. Each scriptural writing is given its own story world, where man is the actual god over God's words:

> If the narrative critic does not think that the story-world corresponds to the real world, then the significance of the narrative for the critic's own life may be minimal or nonexistent. If, on the other hand, the narrative critic (on other grounds) regards the story-world as a faithful representation of historical events (as I do), then the narrative-critical description of the "story-world" becomes normative for life and faith.[23]

It is not Christian to treat the Word of Christ as if it has only the authority of silly children's fables. It is always sin to disregard Scripture's direct authority and teachings—to treat Christ as nothing.

The latest moderns do not like the debate over truth and the negativity of historical criticism, but they reject none of its foundational assumptions. Ahistorical narrative methods neuter all factual claims of Christianity for the ability to tell non-confrontational stories. The Gospel is a story, but one from God that is accurate in all its scriptural details. For example, the virgin birth "is a *biological* truth as well as a *spiritual* one."[24] Exactly like the post-modernist camp, narrative theology is ultra-rationalistic, unwilling to deny any judgment by the critical subject. Narrative critics aim to stay positive (and scholarly) by evading the whole question of truth.

It is reasonable that "reason itself would . . . be subjected to a similar critical appraisal toward the end of the Enlightenment period."[25] The post-modern approach cannot move past atheism because it involves no repentance but a further retreat from facts. "It is the denial of any

---

[22] Gabriel Fackre, *The Doctrine of Revelation: A Narrative Interpretation* (Grand Rapids: Eerdmans, 1997), 10.
[23] J. A. Gibbs, *Jerusalem and Parousia*, 24, 23.
[24] Norman L. Geisler in *Biblical Errancy*, ed. N. L. Geisler, 21.
[25] Alister E. McGrath in *The Science of Theology,* ed. P. Avis, 216.

relationship between religion and matters of empirical knowledge."[26] In modernism man verifies reality, deciding truth by his own powers. In post-modernism this claim is more explicit: man makes reality. The Bible may offer new perspectives, but not truth, because no part of religious truth intersects with scientifically-determined truth.

Despite its grand claims of moving beyond modernism, narrative theology moves further in the same atheistic direction. It does not want to look modern, but wants to retain the modernist divine status seized by man. The whole category of non-factual fiction is a modern one, as the noted writer C. S. Lewis describes: "There are only two views [of the Bible]: either reportage of the facts (history) or the writer suddenly anticipated the whole technique of modern, novelistic, realistic narrative."[27] In the truth of modernism, everything is fiction, even God's revelation, until man verifies it as real and factual by his stringent, scientific criteria.

Exactly like Barthianism, narrative theologians attempt "to move beyond the modern debate of the Bible." "They have moved away from a narrow concentration on whether or not the information in the Bible is historically true."[28] This is also a move away from the incarnation and all of God's involvement with man. If Christ did not die and rise, we are still in our sins (1 Cor. 15:12–19). Saying that Scripture is a resource or horizon-expanding guide is a weak non-confession that even a pagan could make. The facts of Christ's death and resurrection are not needed to maintain such an "authoritative" view of Scripture. All attempts to re-appropriate theological meaning without solidly determining the basis for spiritual and historical facts are damned: "This means that the Holy Bible that is sitting on my desk right now is not directly the Word of God because God is not working through it to reveal himself to me."[29] Accordingly, there is no actual Word of God to debate or teach, only the potential for human words to mimic the ineffable Word of God. Any attempt to hide Christ's words away from criticism is an implicit

---

[26] J. C. Livingston, *Modern Christian Thought*, 38.

[27] C. S. Lewis, "Modern Theology and Biblical Criticism," *Brigham Young University Studies* 9 (1968; https://ojs.lib.byu.edu/spc/index.php/BYUStudies/article/viewFile/4342/3992); reprinted from *Christian Reflections,* ed. Walter Hooper (Grand Rapids: Eerdmans, 1967), 15.

[28] P. H. Nafzger, *These Are Written*, 50, 51.

[29] P. H. Nafzger, *These Are Written*, 51.

denial of Him.

The sinner needs divine doctrine and facts to secure holy faith. "[God] speaks *definitively, ultimately, and decisively,* and *for all time* in his Son, Jesus Christ, the personal Word of God."[30] But what does this eternal Word say, and can it be discussed by man at all? The historicizing of revelation means that Christ does not actually say anything, therefore man is free to speak or do anything as his own god. The "authoritative Christ" is glorified magnificently, but only because all we hear is "ye shall be as gods." The scriptural Christ warns that saying His name repeatedly is insufficient for eternal life: "Not everyone who says to me, 'Lord, Lord,' will enter the kingdom of heaven, but the one who does the will of my Father who is in heaven" (Mt. 7:21). In modernist Christianity ornate motifs and exquisite insights that tickle the mind abound, but the will of Christ is unknowable.

God's will has been revealed in Christ's scriptural words. To put Christianity beyond the reach of historical criticism is simply to give up any possibility of factual truth. The problem with historical criticism is not the "historical" aspect, it is the "critical" part, which leaves man free to criticize all external authorities and establish truth for himself. To say Scripture can be had without concern for its history, makes every phrase of the second article of the Creed—which is true history established *by* Scripture—irrelevant.

Conclusion

The initial excitement over new scriptural methods has clouded the theological shift. Scripture to moderns only contains various human vantage points. The speaking Christ takes a backseat to human opinions. Where men and their authority are trusted, God is not believed. True faith is related to the authority of the divine Word. The bedrock of certainty is believing that Christ still speaks in Scripture and scriptural proclamation. "We censure the doctrines of men not because men have spoken them, but because they are lies and blasphemies against the Scriptures. And the Scriptures, although they too were written by men,

---

[30] P. H. Nafzger, *These Are Written*, 6.

are neither of men nor from men but from God."[31]

No controversy is settled in the LCMS, because there seems to be no divine authority binding us. Everyone has his own foundation, his own truth. Where there is only critique, factions will develop, but they can never be healed on the basis of the truth. Different groups tout their favorite styles of worship, but the divine words are just another tradition. Programs (a more visible type of method) seem more real than scriptural law and Gospel.[32] The appeal to external measurement instead of the Spirit's work through the Word can only mean death to a church. Even if done in the name of Christ and the Gospel, all methods will fail without the actual Word of God bringing the Holy Spirit.

The mere word "scientific" carries more weight than God's Word. People bow to it on the authority of men. Modern theologians like Barth and Bonhoeffer are bandied about as virtuous revelations of truth, more than Scripture itself. Non-confessional, modern theologians are inserted into a confessional framework, destabilizing it. Of Bonhoeffer it has been said: "it is not Jesus Christ, but the Word of God, indeed all religious concepts as such, that he finds problematical. The question of non-religious interpretation derives directly from the foundation and heart of his theology, from his Christology."[33] Because Bonhoeffer "took the closing of the cosmological gaps by science and naturalism as a given," his value can be only of the most limited variety. "He therefore devoted most of his attention to addressing only 'personal facts' of human existence—that is, morals and politics."[34] All is reversed: we critique Scripture and see mere men as infallible. Bonhoeffer wrote: "By all means we must ascertain the fallibility of the [scriptural] texts and thereby recognize the miracle, that we always hear the Word of God from this human word." In the German version of *The Cost of Discipleship*, he denies all factual truth in Christianity:

---

[31] Luther, *Avoiding the Doctrines of Men (1522)*, LW 35:153.

[32] A program cannot be legalistically turned into something else: "One may refer to the *Ablaze!* movement, vision or initiative, but never to the *Ablaze!* 'program'." One may rightly wonder if even Scripture may disagree with synodical authority. The Lutheran Church—Missouri Synod, "The Official Stylebook," (updated Jan. 2016; http://www.lcms.org/Document.fdoc?src=lcm&id=847), 3.

[33] G. Ebeling, *Word and Faith*, 107.

[34] James Harris, *Analytic Philosophy of Religion*, vol. 3 of *Handbook of Contemporary Philosophy of Religion* (Springer Science & Business Media, 2013), 310.

> The confusion of ontological statements with proclaiming testimony is the essence of all fanaticism. The sentence: Christ is risen and present, is the dissolution of the unity of the Scripture if it is ontologically [factually] understood. . . . The sentence: Christ is risen and present, strictly understood only as testimony of Scripture, is true only as the word of Scripture.[35]

So all his nice statements say nothing about the real, sensible world. Therefore, in true modernist fashion, "even holy scripture would have to be contrasted with God's Word" and truth.[36]

Wilhelm Löhe is another theologian, albeit with actual Missouri Synod ties, who seems to exert more influence than the past theologians of the LCMS. While his organizational work is commendable, his flawed doctrine of Scripture changed the nature of his theology. Löhe and other neo-Lutherans,

> attempted to practice "confessional theology," and they did so in a self-conscious, ideological manner. . . . They revived confessional studies, employing an idealistic, historical hermeneutic in which the teachings of the Lutheran Confessions, including the ancient symbols of the Apostles', Nicene, and Athanasian Creeds, were conceived as a development of dogma through which the church experienced both continuity and change in the understanding of revelation across time.[37]

Löhe, for all his outward conservatism, worked with a modern view of the Scriptures, confessions, and truth itself.

Where did this doctrinal development lead Löhe? He developed a different gospel, or rather, two distinct gospels.

> First, in his 'New Aphorisms' of 1851, Löhe declares that there is an essential difference between the word of forgiveness spoken by one Christian to another and the word of forgiveness spoken by the office of the minister: only the latter provides the gift of absolution. Here, Löhe clearly oversteps the bounds of any recognizable Reformation position.

Löhe unfortunately displayed the modernist spirit of historical distance and accommodation to doctrinal change in the name of progress. Scrip-

---

[35] Quoted in: R. Weikart, "The Troubling Truth about Bonhoeffer's Theology," 4–5.
[36] G. Ebeling, *Word and Faith*, 175.
[37] W. Sundberg, "Wilhelm Löhe on Pastoral Office and Liturgy," 192.

ture was insufficient and outmoded for him, at least to some degree. His "theological position, especially on [the] pastoral office and liturgy, bears little resemblance to Lutheran identity as forged in the Reformation. Its usefulness today as an inspiration or guide is also highly questionable."[38] This does not mean we cannot appreciate his practical efforts, but that one should scrutinize his theology in a most critical manner by the scriptural principle.

The theologians that appeal so greatly to conservative Lutherans today—those like Sasse, Löhe, Bonheoffer, and Werner Elert—were all questionable, if not entirely unconfessional, on the basis of theology—God's Word. Citing a carefully turned phrase of a brilliant mind has become more effective than the Bible in the modern LCMS. Conversely, our own tradition, that of Walther and Pieper, is dismissed without even serious consideration. That signals that confessing, in the way the LCMS was founded to do, has changed. We have itching ears to hear something new (2 Tim. 4:3). The Word of God is no longer satisfying or conclusive to our modern, scientifically formed minds.

This view of an LCMS pastor expresses the modernist position well:
> If the Reformation principle *sola Scriptura* was a steadfast anchor for the Reformers, then its misuse or rather purposed perversion through extreme expressions of fundamentalist loyalty to the text of the Bible, verbal plenary all-authoritative inspiration, and inerrancy, has become a crippling intellectual impediment to the gospel, and it does so in this way: When the natural, historical and theologically legitimate boundary providing extra-biblical means of authority are eliminated and the Bible is touted and used as the *only* means of authority (true to foundationalist principles) and issues of geography, marriage, geology, economics, politics, and matters secular are elevated to the same level as that of the metanarrative of Scripture, as that of grace, faith, and the gospel, then the material authority of the Word in the Bible is undermined. The gospel is relativized and trivialized when set on equal footing as biblical business principles. . . . The gospel message has been embarrassingly squelched by tens of thousands of interpretations, all equally authoritative, all standing on equal authority.[39]

---

[38] W. Sundberg, "Wilhelm Löhe on Pastoral Office and Liturgy," 194, 191.

[39] J. Bombaro, "Biblicism," in *Built on the Foundation,* ed. T. Simojoki, 165.

Disconnected from the Bible and the physical, factual world, the Gospel is without authority—it becomes merely a mental construct. If the biblical manger of the Gospel is an embarrassment, the biblical Christ is also likely to be. Christ spoke not just inaccessible words about intangible things, but about the creation and people He made. To limit the authority of the Bible is to limit the authority of Christ. Luther spoke of Satan's work in drawing people *away* from the Bible to a mythical gospel that has no connections to real life:

> In short, the devil is too clever and too mighty for us. He resists and hinders us at every point. When we wish to deal with Scripture, he stirs up so much dissension and quarreling over it that we lose our interest in it and become reluctant to trust it. We must forever be scuffling and wrestling with him. If we wish to stand upon the councils and counsels of men, we lose the Scriptures altogether and remain in the devil's possession body and soul.[40]

At least those in the Devil's control have the consolation that they are not fundamentalists.

Real science remains a mystery to those devoted to the cult of scientism. They accept science in the most unscientific fashion. Theologians who cannot duplicate hard science and do not understand the philosophical basis of scientific "facts," are all too happy to believe without evidence. "It is nonscientific to accept science only on authority."[41] When non-scientists accept scientific conclusions on the authority of scientists, they do so on the basis of faith and external authority. Man's word is trusted more than God's Word. The true theological scientist must say that "death is the boundary of historical statements. For the rest, the modern view of history cannot make an exception even in the case of the historical Jesus."[42] All witnesses, even Christ Himself, must submit to extreme doubt before they are dignified with the authority of telling historical facts.

Theology, unmoored from Scripture, becomes an exercise in novelty. Perspectives, insights, and fads replace the conviction which comes from having the truth. "All talk of commitment to the Confessions is senseless when Scripture is lost as concrete judge over all proclamation

---

[40] *That These Words of Christ, "This is My Body," etc., Still Stand Firm against the Fanatics* (1527), LW 37:17.

[41] C. Sagan, *The Demon-Haunted World*, 251.

[42] G. Ebeling, *Word and Faith*, 292.

and doctrine."[43] All things get caught up in the historical stream of critical relativism. Divine authority is lost without the doctrine of inspiration. The ability to condemn on the basis of God's Word is crucial to a confessing church body—that is, one that actually says something. This is why "the condemnatory judgments of Luther and the Lutheran Church are always accompanied by reference to Bible passages."[44] Only divine authority can be appealed to when men differ and disagree. Only God conclusively settles a dispute.

Where Scripture is not the practical norm of truth, God is assumed to speak less clearly than man. Without a divine authority binding all, unity will be impossible, and then church becomes an excuse for all-inclusive acceptance. The church without God's literal Word will have no divine conviction, even when it speaks something worthwhile. Eventually, it will have nothing worth saying.

While the conclusions of recent critical exegesis in the LCMS have mostly appeared to be within traditional doctrinal lines, doctrine has been severed from the foundation of Scripture. The separation of theology proper from exegesis is highly problematic for a scriptural church. It is easy for the LCMS to defend tradition for the sake of tradition, but not from Scripture.[45] We have lost the conviction of divine authority, and therefore the ability to confess.

The glory of confessional theology is that the truth is known and fixed in specific writings.[46] This certain, divine prejudgment is the antithesis of scientific autonomy. To say "God says thus," and call the opposite sin against the Holy Lord, is the basis for true Christian unity. Critical methods will only lead to the fragmentation and disintegration of a church body. This gradual erosion means unity will be impossible. Tradition alone is not enough to maintain a church that cannot make judgments based on God's Word.

C. F. W. Walther's words are a stern warning for those in the Synod which he guided at its inception:

---

[43] Peter Brunner, "Commitment to the Lutheran Confession—What Does It Mean Today?," *Springfielder* 33:3 (1969), 4–5.

[44] H. Gensichen, *We Condemn*, ix.

[45] Examples abound in the realm of Communion practices.

[46] H. Berkhof, *Two Hundred Years of Theology*, 62.

as often as we open our Bible, we intend to think of the word of the prophet Isaiah: "Hear, O heavens, and give ear, O earth, for the Lord speaks." That should be our emblem on the battlefield, we should emblazon these words on our banner, and we trust alone under this sign for victory. Surely, if ever our synod should no longer hold this banner aloft, then her fall would not merely be imminent but would have already taken place. Then it is far out, that it should be thrown out from this earth like the insipid salt; it deserves to be trampled by the people (Mt. 5:13).[47]

---

[47]C. F. W. Walther, "Evening Lectures on Inspiration," Lecture IV.

Part IX

A Foundation for Confessing

Chapter 31

Timeless Truth

The only godly approach is to take Christ's Word seriously and submit all human thoughts to it. "So everyone who acknowledges me before men, I also will acknowledge before my Father who is in heaven" (Mt. 10:32). Confessing propositions—that is factual, everyday knowledge—about Christ is essential to being a Christian. It is not an academic extra, but a result of living in Christ, possessing His Spirit, and being justified by the Father.

Swallowing 7,000 Bibles will not make one a Christian. Neither does the confession of Scripture's inerrancy, but it cannot be denied without the severest consequences. Inerrancy is a property of God and His Word to be used to proclaim the inerrant truth of Christ. Who would confess an errant god or one who requires men to correct his writing?

Lutherans have the strongest confession of God's Word—not a slogan, but the sum of its saving content in written form. The Book of Concord *is* the Word of God, because its words correctly summarize and agree with the Scriptures. Only because its teachings are derived correctly from Scripture can these historical documents be a present norm—not just a historical remembrance of past controversies. Pure doctrine, with *no error*, defines the true Lutheran church. She claims a doctrinally inerrant, though not complete, confession of God's Word. The inspiration of Scripture allows a "direct identification of theology and truth, as long as the former faithfully reflects the Word of God."[1]

---

[1] Hendrik Krabbendam in *Challenges to Inerrancy*, eds. G. R. Lewis and B. A.

This is no abstraction, but the basis for the surety of Christian doctrine.

> One is therefore infallible and free from error in all things when one clings to the Word of God. For as sure as the Bible is the Word of God and inspired by the Holy Spirit, as sure as Christ is the Son of God and the mouth of the eternal truth, so certain is it also that we cannot err, if we hold onto the letter of Holy Scripture.[2]

The doctrine of Scripture is no theory or theological question. It is part of the unified truth, revealed by the Spirit, which testifies to Christ. To lose inspiration, means inevitably losing the scriptural Christ.

The modernist thinks in terms of all authority being centered on man. Christian orthodoxy, despite some disagreement over the means and location of the divine teaching, sees truth as having divine authority over all earthly methods and opinions. Fundamentalism has been defined as "the demand for a strict adherence to certain theological doctrines."[3] This is not a knee-jerk reaction to modernism, as is claimed. It is a rejection of a new error (critical modernism) by the truth which has always been the supreme truth.

Orthodoxy does not see Scripture as a prison, but the only source of the freeing truth. Christ does not put a heavy yoke on those who submit, but submission is impossible without faith in the living Jesus. At its widest definition, fundamentalism simply means to hold to an unquestioned standard outside of man's reason. It is to reject the mental god of man's intellect, though not necessarily to confess the true God. It is not to be an intellectual atheist, since the very concept of divine revelation rules out the sufficiency of scientific verdicts.

The burden of pastors, and hopefully professors, is to say something theological—to speak for God. Unless man is simply identified with the deity, this is not an easy task. However, sinners are still called by Christ to preach, so they must have a source and norm for their words. Modern theology works with the assumption that "a mere refurbishing and repetition of the theology of the reformers is . . . utterly impos-

---

Demarest, 315.

[2]C. F. W. Walther, "Communion Fellowship (1870)," in *Walther's Works: Church Fellowship*, 151.

[3]"Fundamentalism," *Wikipedia, The Free Encyclopedia*, https://en.wikipedia.org/w/index.php?title=Fundamentalism.

sible."[4] When the actual words of God are denied, the truth will be impossible to preach or confess. Then dogmatics will be "the attempt to let the unobjectionable truth mirror itself in the waters of scientific objectifiability."[5]

God does not change, and neither does His truth.

> Remember your leaders, those who spoke to you the word of God. Consider the outcome of their way of life, and imitate their faith. Jesus Christ is the same yesterday and today and forever. Do not be led away by diverse and strange teachings, . . . which have not benefited those devoted to them (Heb. 13:7–9).

Speech from God is deemed unscientific by moderns. Scripture, and therefore Christianity, is in a Babylonian captivity of the scientific mindset.

To speak, believe, or preach "thus saith the Lord," one must first have His actual words. One book summarizes all of 20th-century theology as the focus on this one question: "in what sense does God's voice collide with this world?"[6] Barth's view, that the Bible is not the Word of God apart from an unknowable, capricious act of the Spirit, means that "the pastor can never be sure whether the proclaimed word was [his] word or God's Word."[7]

Without a sure, revealed word, "no theological system could claim to have captured the Word of God in full."[8] Scholarly doubt, first introduced as the leaven of scientific study, undermines the whole foundation of Christianity. Like the prophet Elijah, Walther taunts the false theologians of modernism: "Let them maintain their Bible, if they can die on the basis of it. We don't want it."[9] A teaching of Christ that does not give confidence to die for the truth is not worthy of the name Christian. In the scriptural Christ, "we have boldness and access with confidence through our faith in him" (Eph. 3:12).

What happens when theology is not based on divine revelation? "No form of preaching nor theological system can claim timeless validity and

---

[4] G. Ebeling, *Word and Faith*, 18.
[5] H. Berkhof, *Two Hundred Years of Theology*, 80.
[6] S. J. Grenz and R. E. Olson, *20th-Century Theology*, 312.
[7] U. C. Scharf, *The Paradoxical Breakthrough of Revelation*, 81.
[8] U. C. Scharf, *The Paradoxical Breakthrough of Revelation*, 87.
[9] C. F. W. Walther, "Evening Lectures on Inspiration," Lecture XIII.

simply be accepted by later generations."[10] Inerrancy is a narrow issue, but the error behind its denial threatens to overturn all of theology. This trait of inerrancy has always been held by Christians, and it must be mightily defended against rationalist approaches to truth. Even a less-than-fundamental error can wreck the entire theological enterprise, leaving no truth left to be believed.

At the heart of modernist thinking is doubt that the truth can be condensed or expressed in any creed or confession. Do "the confessions contain *a* true exposition of the Bible, but not *the* true exposition?"[11] Only words having God's own authority can grant divine faith. Without divine authority, there is only human uncertainty. Luther addressed this very modern issue:

> [hypothetically,] if the Word finds not the spirit, but an ungodly person, then it is not God's Word, thus defining and fixing the Word, not according to God, who speaks it, but according as people entertain and receive it. . . . A true Christian must hold for certain that the Word that is delivered and preached to the wicked, the dissemblers, and the ungodly, is as much God's Word, and that the true Christian Church is among sinners, where good and bad are mingled together.[12]

Neo-orthodoxy has a "symbolic truth, not a literal truth."[13] "[Creeds] are no longer understood as facts, but merely as [figurative] symbols . . . ." "They cannot be reconciled with the premises governing our scientific and technological mastery of the world."[14] The truth is not saved by denying it. This giving away of facts is placing the truth of God in Moses' basket for someone else to confess.

All talk of the dynamic power of God's Word without solid and real truth is a sham. One should much rather be silent than incur God's wrath, if this be the operative method. "The conception of the Word as the norm of truth and doctrine is inseparably united with that of the Word as a means of grace."[15] At stake "is the task of preaching and

---

[10] H. Thielicke, *Prolegomena*, 25.

[11] Harold Ditmanson in *Studies in Lutheran Hermeneutics*, ed. J. Reumann, 92.

[12] Luther's *Table Talk*. R. Comfort, *Luther Gold*, 94.

[13] M. Halverson and A. Cohen, quoted in: *Studies in Lutheran Hermeneutics,* ed. J. Reumann, 325.

[14] Helmut Thielicke, *How Modern Should Theology Be?*, trans. H. G. Anderson (Philadelphia: Fortress Press, 1969), 6.

[15] J. Köstlin, *The Theology of Luther*, 2:215.

teaching in the church and to the world."[16] The denial of revelation is ultimately a denial of Christ, for "we know nothing of Jesus except through His words."[17] "To recognize and accept a revelation means quite simply to capitulate to it unconditionally, to surrender everything that belongs to the nature of the godless world" and to "let go of the whole superstition that calls itself science, above all, historical science."[18]

The Word of God is not just a book, it is the norm and source of all truth that God has given. An appeal to a non-scriptural Christ is an appeal to a false spirit. The true Christ is not locked away in the past—He speaks today in the writings of Moses, the prophets, and the apostles. A true Lutheran must confess inspiration, so that the Lutheran Confessions are a pure and accurate rendering of the actual Word of God, as they claim to be: "the pure doctrine is with us."[19] This Christian conception of Scripture does not cause one to bow down to the letters and words, but it treats it like all other human communication that is understood according to its grammar. Its teaching is to be promulgated and spread, so the Spirit can illumine the human mind and turn men to Jesus in faith. "Essentially God's Word is not the letters and words but the divinely revealed meaning."[20] The letters and words are how the revelation is communicated, so the inspired words are the only way to access the inspired meaning. The truth of Christ is one, so all doctrines are organically connected, and bring us to Christ. This single body of truth originates from Christ, through the Spirit, and leads to Him. Christ said: if you continue in my Word (not my bare, unknowable essence), then you will know the truth (Jn. 8:31).

There is no eternal, inaccessible word or gospel behind the accessible one. To safeguard the Bible from criticism, the very Word of God must be banished.[21] There is only one Gospel, even if all the angels sing their critical consensus about its changing nature—let them be accursed (Gal. 1:6–9). This truth is testified to in every statement of true confessional theology, though we cannot expect the Confessions to explicitly address a problem which did not yet exist. Man cannot confess the majesty

---

[16] Samuel Laeuchli, quoted in: J. W. Montgomery, *Crisis in Lutheran Theology*, 45.
[17] H. D. McDonald, *Theories of Revelation*, 2:333.
[18] Erwin Reisner, quoted in: G. Ebeling, *Word and Faith*, 18.
[19] FC SD Preface, 8.
[20] A. Hoenecke, *Evangelical Lutheran Dogmatics*, 1:458.
[21] G. Maier, *End of the Historical-Critical Method*, 25.

of Christ by his own sin-bound powers. One can only speculate on it, as an intellectual work, apart from knowing God's revealed will. "The Confessional principal presupposes the Biblical, since Reformation doctrine rests squarely on Scripture alone." Without a foundation, the Gospel dissolves into a "short slogan," or "a cut-rate, mini-gospel," instead of the body of doctrine revealed by Christ.[22] Scripture protects and safeguards the Gospel.

Truth touches men and binds them to God in the form of doctrine. While non-doctrinal approaches based on man must change and compromise, God's Word is timeless. Man-centered truth results in absurd statements, like: "certain interpretations of 'Thou shalt not kill' can certainly inflict great suffering on others." God says that all violations of His commands inflict suffering on His Word, so He is justly angered. If man determines truth, "in principle ethical questions cannot be answered from the Bible alone."[23] Once God's Word is said to be only man's limited word, it does not appear to speak clearly on anything. Man has refused to listen to anything but himself, and as punishment, God obliges. "To Barth the important thing is that God speaks. It does not matter so much what he says, because God's speaking is itself his self-revelation and reconciliation."[24] A wordless god is the true gospel to moderns.

Propositional truth gives finality and authority to human formulations of doctrine. An old Eastern Orthodox confession begins the topic of revelation thus: "Whence is the doctrine of the orthodox faith derived? From divine revelation."[25] Yes, this church incorrectly confesses two channels of revelation, like the Roman church, yet both sources are said to come from God's Spirit, and, therefore, are inerrant and propositional. Only certain propositions can support the glorious Gospel of Christ, which gives hope in a dying world. The concern here is not just individual passages, but the divine doctrine they contain. One who cites more Scripture is not more faithful. It is not a mechanical authority but the living oracles of God.

---

[22] K. E. Marquart, *Anatomy of an Explosion*, 6–7, 18.
[23] J. Barr, *Beyond Fundamentalism*, 123, 122.
[24] P. H. Nafzger, *These Are Written*, 123.
[25] *The Longer Catechism of the Orthodox, Catholic, Eastern Church* (Moscow, 1839). P. Schaff, *Creeds of Christendom*, 2:446.

It is a sad irony that most scholars of Martin Luther's theology want to sound like Luther, but deny the reason he was able to stand firm against the world and the Devil himself. Luther emulated Christ, and acknowledged "It is written (by God)" for every word of the Bible. God's Word animated Luther to such a high degree, that his theology is simply God's theology. Not by revelation, but due to the propositional nature of God's Word. The Spirit moves man to appropriate divine teachings and proclaim them. Paul does the same when he most boldly calls Christ's gospel, "my gospel" (Rom. 2:16, 16:25; 2 Tim. 2:8). This is a spiritual exercise: to live in Christ and speak for Him. "I have been crucified with Christ. It is no longer I who live, but Christ who lives in me" (Gal. 2:20). It is certainly not a Spirit-less, academic endeavor to be in the living God by faith.

Scripture gives the pure truth, and this truth can be handled by man. "The Evangelical Lutheran Church is sure that the teaching contained in its Symbols is God's pure truth because it agrees with the written Word of God in all points."[26] Scripture contains propositional teaching, which allows it to be restated in confessional form. This is not said to be easy for sinners, but statements can be compared to Scripture. We are to use the divine writing to verify human writing. "We are sure of our doctrine, because it is taken from Scripture."[27]

To believe, as the object of faith, that man's word is God's Word is blasphemy. It is the very definition of idolatry. Faith is in the Word of Christ (Rom. 10:17). That does not mean repeating syllables of Scripture, but that Scripture can be used so that a sermon based on the divine Word is also a divine word, in a derivative sense. It is not a norm, but it is just as valid for sustaining faith, if it does not deviate from the sound words of the Spirit. "You have searched the Scriptures, which are true, which were given by the Holy Spirit; you know that nothing unrighteous or counterfeit is written in them."[28] Proclamation is measured by Scripture, not by anything else.

Christianity touches every area of life because it deals with Christ,

---

[26] Thesis XXI. C. F. W. Walther, "The True Visible Church of God on Earth," in *Walther and the Church*.

[27] T. Engelder, *Haec Dixit Dominus*, 5.

[28] Ignatius of Antioch, *Letter of the Romans to the Corinthians [1 Clement]* (A. D. 95–97), 45:2–3. *The Apostolic Fathers*, ed. M. W. Holmes, 79.

through whom all things were created. Elements of Christianity might be intellectually fulfilling, but Christ commanded actions. The Lord's Supper is something to be taken in the mouth, not to be adored with the intellect. Private prayer is not solely a thought process within the mind, but a movement of the Spirit, who cries "Abba, Father" within us (Gal. 4:6). Preaching is the communication of God's Word, in the sight of Christ, to sinners with divine authority, not an individualistic rationalizing. "For we are not, like so many, peddlers of God's word, but as men of sincerity, as commissioned by God, in the sight of God we speak in Christ" (2 Cor. 2:17).

The lifeblood of the Church is the truth of Christ. It is to be preached without apology to the world. To fulfill this call is to speak for the living Christ, even if it means being abused like Him. Those who embrace the world's idea of the unreality of theology will find out too late that Luther was actually right: "The God of the law, the Last Judgment, and hell were real in Luther's day . . . ."[29]

---

[29] H. Thielicke, *How Modern Should Theology Be?*, 10.

Chapter 32

Meaningless Confessional Subscriptions

The mere fact that the early Lutherans made a unanimous confession is a powerful testimony to their doctrine of Scripture. Why has there been so little confessing since 1580? Heresies have not ceased and the Spirit has not retired to the old folks' home. Historical criticism is the acid which dissolves confessionalism, since doubt is not compatible with a believing heart and a confessing mouth (Rom. 10:8–10). "It is obvious that with this [historical-critical] approach there can be no such thing as a *quia*[1] subscription to the Lutheran Confessions."[2] Without a single divine fountain and norm of truth, the attempt to reproduce the truth is a fantasy.

"Mere assertions of the Bible's authority mean little unless the Bible's actual substance is confessed."[3] What defines a confessional Lutheran is not a confession *about* Scripture, but that the Bible's teachings *are* sufficiently communicated in specific human documents: the Lutheran Confessions. A human confession is believed to be the divine truth of Christ. Without a clear authoritative source and norm, it would be folly to say that a human writing could contain enough truth to be binding on man. "Confessing the confessions does not mean mere subscription,

---

[1] That is, an unconditional subscription to the human writings *because* the writings agree with God's Word.

[2] E. Moeller, "The Meaning of Confessional Subscription," 200.

[3] K. E. Marquart, *Anatomy of an Explosion*, 7.

but a commitment to the truth of the Word of God and the person of God who speaks the Word."[4]

Ecumenism, inclusiveness, and modernism go hand in hand. When the sole authority of Scripture is lost, there is no longer a criterion to distinguish between churches. "All talk of confessional allegiance is meaningless, if Holy Scripture is lost as the concrete judge over all proclamation and teaching."[5] The confessions are nothing but an application and interpretation of Scripture. "*Sola Scriptura* is written on every page of the Lutheran Confessions."[6] The scientific relativizing of truth means that the "old debate about *quia* and *quatenus* [insofar as] are at bottom irrelevant."[7]

The 16th-century confessional documents do not pass for scholarly exegesis today, though.

> In the era of historical critical research the strict theory of inspiration has become an impossibility. Consequently, those Fundamentalists who cling to the theory of verbal inspiration of the 17th century exclude themselves from the responsible ecumenical discussion on the concept of biblical inspiration.[8]

Divine certainty, which means rejecting the papacy of scholars, is forbidden. Not individual doctrines, but the mere possibility of a "perennial theology in the Book of Concord" is questioned.[9] Yet the LCMS was built on having God's Word, not just in Scripture, but in its confessions and in its pulpits. The strongest statement of the divine inspiration of Scripture is not made concerning Scripture itself. The confessions of the Lutheran Church are God's Word, not in the sense of the individual words (as with Scripture), but that what they actually teach is God's pure truth. All solid confession of the truth is founded on a certain, understandable revelation. God has chosen to speak only in Scripture to this adulterous generation: "They have Moses and the Prophets; let

---

[4]Lewis W. Spitz, "Discord, Dialogue, and Concord: The Lutheran Reformation's Formula of Concord," *CTQ* 43:3 (1979), 191.

[5]Peter Brunner, quoted in: K. E. Marquart, *Anatomy of an Explosion*, 35.

[6]Eugene F. Klug, "Discussion Outline on the *Sola Scriptura* Essay," Faculty study paper (Copy in Concordia Theological Seminary Library, Fort Wayne, IN, 1968), 1–2.

[7]G. Ebeling, *Word and Faith*, 177.

[8]M. Ruokanen, *Doctrina divinitus inspirata*, 145.

[9]M. R. Noland, "Walther and the Revival of Confessional Lutheranism," 213.

them hear them" (Lk. 16:29).

Inspiration means that Scripture is no academic matter. To have God's Word and treat it as non-authoritative, human words is the sin of unbelief. "On Judgment Day these words will accuse all those who have robbed Holy Scripture of the privilege of speaking with words which the Holy Spirit teaches:"[10] "Now we have received not the spirit of the world, but the Spirit who is from God, that we might understand the things freely given us by God. And we impart this in words not taught by human wisdom but taught by the Spirit, interpreting spiritual truths to those who are spiritual" (1 Cor. 2:12–13). All confessions (even the unintentional ones) of God's Word are made before the true Critic, who will judge the living and the dead. They are spoken before Christ in view of Judgment Day. Luther powerfully describes the Last Day on the basis of Mt. 25:31–46: "Dear sir, listen, you have also pretended to be a Christian and boasted of the Gospel; did you not also hear this sermon, that I myself preached, in which I told you what my verdict and decision would be: 'Depart from me, ye cursed'?"[11] The words of Scripture and faithful scriptural sermons give the verdict in advance.

The type of confessional subscription made shows if one takes the Scriptures seriously. "From the attitude which one takes toward the Symbols of the Lutheran Church we learn whether he knows and accepts the scriptural doctrine."[12] If one denies verbal inspiration, in the neo-orthodox way, then "divine revelation is never the communicating of truth."[13] Without truth, theology is a poor orphan without a voice. The Church should be bold: it has a hope to which science can never attain. "Nothing is better known or more common among Christians than assertion. Take away assertions and you take away Christianity."[14] Assertions are propositions which claim to be inerrant, that is, completely true.

The most harmful error might be to think that the old Lutherans did all the hard work in defending the truth. Many think that all theological problems are now solved, so the truth will defend itself and be rationally convincing, but the Confessions, while true, are not exhaustive. They

---

[10] C. F. W. Walther, "Evening Lectures on Inspiration," Lecture VIII.
[11] *Complete Sermons of Martin Luther*, 3.1:385–386.
[12] F. Pieper, *Christian Dogmatics*, 1:358.
[13] K. Kantzer, "Revelation and Inspiration in Neo-Orthodox Theology," 222.
[14] Luther, *The Bondage of the Will* (1525), LW 33:21.

state enough propositions to rule out the doctrinal aberrations their authors encountered. They framed the divine truth so that the scriptural teaching would not be perverted by Satan's then-current assaults.[15] Yet in no way do they replace Scripture, claim to be a revelation, or mean that future confessing is not needed.

The Church is always in a battle for the truth. One's view of Scripture impacts one's confession of Christ—the very core of Christianity. "It had been assumed for centuries that the ecclesiastical creeds had given the final and unalterable answer, and whoever did not believe them could not be saved."[16] This relativism, based on a prolonged attack on external authority, is itself anti-Christian. The modernist, in trying to describe fundamentalism, is at a loss. To the mental idolater, it "is nothing more than the absence of irony. [Moderns] cannot get past the essential literalness of the believer."[17] The fundamentalist claims a truth above all men—an incredible thing to one so steeped in method that he cannot think without it.

> The Formula of Concord was only possible because there was a common understanding of the identity, the meaning of, and the authority of Scripture. . . . I am convinced that unless and until Synod, or an overwhelming majority of its members, arrives at a consensus on the doctrine of Scripture, Synod will more and more become a divided camp, doctrinally speaking."[18]

We must get over the childish idea that spouting mere theological formulas about inspiration or inerrancy is enough, while God's words play no vital role in our thinking. It is sinful to treat God's Word worse than a baby's babbling.

What is contrary to the Lutheran Confessions is contrary to the Word of God.

> The truth is that *correct human explanations* of Scripture doctrine are Scriptural doctrine, for they are simply the statement of the same truth in different words. These words are not *in themselves* as clear and as good as the Scripture terms, but as those who use them can absolutely fix the sense of their own phraseology by a direct and infallible testimony, the human words may more perfectly exclude

---

[15] L. Spitz, "Discord, Dialogue, and Concord," 191.
[16] Gerhard Ebeling, quoted in: C. E. Braaten, *History and Hermeneutics*, 54.
[17] Leon Wieseltier in *The Fundamentalist Phenomenon,* ed. N. Cohen, 192–93.
[18] E. Moeller, "The Meaning of Confessional Subscription," 209.

heresy than the divine words do.[19]

Confessions serve mainly a negative function by defending Scripture and leading to it. Its human words are not a replacement for the divine words. There can be certainty about our knowledge of God, though, due to Scripture being the underlying rock and basis. This is why one's exegetical method is deadly serious. "Imitating or guessing is not to be allowed in the explanation of Scripture; but one should and must be sure and firm."[20] Without Scripture there is no basis for factual truth.

Men do not speak better than God. Man's confessing delimits the truth by ruling out false doctrine. It is polemical by nature. Polemics, the condemning of falsehood, is a lost art today. It has receded under the modernist spirit. The peace of atheistic rationalism is much more appealing to the flesh than the bitter polemics and outward divisions of dogmatic orthodoxy. The Gospel as theological "norm" virtually rules out the possibility of divine criticism. It is a license for unlimited rationalism.

Dogmatic judgment is God's own judgment. Doctrine exercises authority over all men and all their thoughts. Propositional doctrine is fundamentalistic, or even terroristic, today, because it destroys the rational basis for the world's peace. It is perceived as hate and violence to modernists, because authority *is* the evil to be rent asunder from man. To be a civilized person is to accept the basic Enlightenment assumptions of personal autonomy and freedom from external regulation. It is to be as God in one's own mind.

One with the truth cannot be silent when it is attacked. The Spirit will move the believer to confess, even if it means persecution. If Christ does not return first, the great issue of this period will eventually be dealt with—by confessing the truth of Scripture.

> As the Patristic age faced a *christological* watershed, as the Medieval and Reformation churches confronted *soteriological* crises, so the contemporary Church finds itself grappling with the great *epistemological* question in Christian dogmatics.[21]

---

[19] Charles P. Krauth, *The Conservative Reformation and Its Theology* (Philadelphia: General Council Publication Board, 1871; reprint, Minneapolis: Augsburg, 1963), 184.

[20] Luther, *Complete Sermons of Martin Luther*, 1.2:101.

[21] J. W. Montgomery, *Crisis in Lutheran Theology*, 15–16.

A Foundation for Confessing

Questions of knowledge and authority can only be answered definitely by submitting even our best and highest thoughts to the simple words of the Holy Ghost, the source of all divine truth.

Among the theological children of Luther, it must be asked "why has the Church remained silent . . . since 1580," or are there not great errors today which undermine the Gospel and demand a counterattack?[22] There are only no heresies to polemicize against if one believes there is no eternal Gospel of the truth of Christ.

---

[22] P. Brunner, "Commitment to the Lutheran Confession," 7.

Chapter 33

Male/Female Roles Are Divine

The vital issues of theology are not abstract, but practical. Sexuality, marriage, abortion, and whether only men may hold the pastoral office are highly controverted issues. These topics are dominant because it is in the body that concrete authority is denied. Authority becomes personal because God in Scripture denies our own authority and freedom to act as we please. He claims absolute authority over all He has made.

In olden days transgressing a single verse of God's Word, no matter the reason, was called sin. Today, morality is where noetic idolatry and Scripture clash most ferociously. God is not a mental idol if He legislates how we are to live in this world. It is in the realm of the sixth commandment on marriage that authority most confronts atheistic man. How we live is addressed by the same Christ who gave His life for the world. How we are made and live, as either male and female, is directly connected to the Redeemer. The scriptural Redeemer is also the scriptural Creator.

"The modern atheistic man, the man who mistakenly imagines himself to be ignorant of God, autonomous, self-reliant, [creates] his own good and evil."[1] The practical atheist can go to church and be involved in churchly activities, but lives as if there is no God. No authority, in his mind, hovers over his body and behaviors. Everyone's favorite modern

---

[1] A. Richardson, *The Bible in the Age of Science*, 92–93.

theologian, Bonhoeffer, wrote in 1929: "The New Testament contains no ethical precept which we may or even can adopt literally."[2] The scriptural and scientific accountings of truth conflict most in standards for sexual behavior. Authority finds concrete grounding in the sexually differentiated bodies Christ made.

"Natural law," in the singular, is a body of knowledge originating from God, but knowable without special revelation. It is revealed by God, nonetheless, and is just as divine as Scripture. The Author and authority of general and special revelation are the same. The knowledge found in both is completely harmonious. "Natural laws," plural, are rationalistic, man-centered explanations which exclude divine cause and authority. They are based on man's sinful observations of the workings of a world in bondage to decay (Rom. 8:21). It is "the conception of natural law which gives us our basis of the [modern] understanding of the world."[3] It allows the world and human life to be discussed without talking about a deity.

Natural law is completely compatible with Scripture and carries the same divine authority. Both define God's will for male and female timelessly and universally. The body and societal relationships are the predominant battlefield of the modernist heresy. Here God's authority practically conflicts with Cartesian autonomy. "I am much more aware now than I was in 1974 about how pervasive patriarchy was in the biblical world and the biblical text. It colors its view of the human situation and of God."[4] This Seminex professor rightly notes that the doctrine of man is linked to the doctrine of God. Human authority structures cannot be overturned with rejecting Christ's authority. Patriarchy—that men have the responsibility of leadership—reflects the authority instituted by God. According to Christ's Scripture: "The head of every man is Christ, the head of a wife is her husband, and the head of Christ is God" (1 Cor. 11:3).

The results of historical science are not neutral according to Scripture. Our "history is essentially that of emancipation of humankind from

---

[2] It is no shock that this leads to a denial of Christ's atonement: "During his time in prison, Bonhoeffer complained that the New Testament was too overgrown with 'redemption myths'." R. Weikart, "The Troubling Truth about Bonhoeffer's Theology," 5–6.

[3] J. Y. Simpson, *Landmarks in the Struggle between Science and Religion*, 161.

[4] R. Klein, "How My Mind Has Changed," 1–2.

patriarchal and authoritarian conceptions of life and society."[5] Its results are visible, thereby boosting the status of the sciences. Everyone is said to be equal, because we have believed the lie: "ye shall be as gods, critiquing good and evil" (Gen. 3:5). Scripture demands that all people submit, even mistreated children, slaves, and wives, because all authority is instituted by God (Rom. 13:1). Submission, for the Christian, flows from having the true God, but the authority God exerts is still valid for the unbeliever—even apart from the Gospel. The scriptural Gospel does not undo any physical orders or authority structures of this world.

The modern destruction of earthly authority alters the doctrine of God. He is not an idea, but the animator and judge of heaven and earth. What God does not condemn, He cannot die for or forgive. Not all authority is the Gospel, but without authority there is no sin or forgiveness of sins. "The anarchist spirit penetrated very deeply into modern culture."[6] In the name of science, freedom from God's authority is progress. God's role in the world is denied under the satanic guise of objectivity. But the Gospel, as much as people might hate the notion, makes no sense without real authority.

The scientific mindset is assumed to have removed the stench of medieval philosophy and pre-modern savagery. The suppositions behind scholarly objectivity, though, are much more arrogant. The radical principles behind rational, scientific criticism have borne their fruit just recently. The ideas of autonomous criticism and the method of universal doubt have paved the way for the acceptance of homosexual actions and transgenderism. It is on social and moral issues, the empirical realities of the body, that modernism has set it sights. "This culture after all rests on presuppositions which in the Old and New Testament are simply described as sinful."[7]

Christ sets straight the marriage issue by citing a very old document: "Have you not read that he who created them from the beginning made them male and female?" (Mt. 19:4). Natural law in creation has no less authority than Scripture. Christ, the Maker, does not contradict Himself. To reject the authority of God in creation is to reject the Lord who entered creation.

---

[5] W. A. Visser 't Hooft, *The Fatherhood of God in an Age of Emancipation*, ix.
[6] W. A. Visser 't Hooft, *The Fatherhood of God in an Age of Emancipation*, 104.
[7] H. Berkhof, *Two Hundred Years of Theology*, 302.

Female pastors are demanded by the rational principle of equality in Enlightenment thinking. It is the continuation of the rejection of external authority. However, Scripture forbids it: "I do not permit a woman to teach or to exercise authority over a man; rather, she is to remain quiet" (1 Tim. 2:12). Scripture does not provide a rational reason, or connect this truth to the Gospel in a logical way. Instead, God simply speaks with authority.

It sounds legalistic and un-Christological to apply old, seemingly outdated standards to today. We have bought the lie that everything has a "historically conditioned form." But Christ, who made man and woman, also speaks in Scripture.[8] He did not approve of every then-current practice. His Word, which gives life, also has authority over social relationships. Submitting to Christ in our bodies is part and parcel of being Christian. "If the criterion of true and false doctrine remains anchored in the article of justification, the yardstick of morality is eliminated."[9] Unrepentant sexual immorality disqualifies one for the kingdom of God and drives out the Spirit (Gal. 5:19–21; 1 Cor. 6:9–11). Christians can only live by the Spirit according to the authority with which He created us. The same Spirit now speaks to us in Scripture with the authority of Christ. If sins are not bound and condemned with divine, unchanging authority, they cannot be forgiven.

It is true that there is nothing specifically Christian about male and female roles, or even marriage. This knowledge which is physically and authoritatively revealed in our created bodies, though, is being denied. In the name of equality the most obvious authority structures inherent between the sexes are abolished. "In the established orthodoxies I know, God has commanded or God's appointed authorities have long decreed that religious truth is reserved to males."[10] The rebellion against authority has now attacked marriage, a true natural and divine law.

Biology and hard scientific data tell us that God made men and women to do different things. Men do not excel in breastfeeding infants, and women do not win "strongest man" competitions. Rebellion against what we were created to be is a rebellion against Christ, the Creator. Even down to the tone of their voices, one carries more *authority* than

---

[8] H. Thielicke, *Prolegomena*, 25.
[9] H. Gensichen, *We Condemn*, 50.
[10] Eugene B. Borowitz in *The Fundamentalist Phenomenon,* ed. N. Cohen, 238.

the other. Man and woman, if externally equal, must clash and be incompatible. For moderns marriage is about internal passions and rationally perceived compatibility—not divinely fixed and authoritative roles.

Female clergy have become the litmus test for modernism. This is the most practical, visible denial of authority, especially of religious authority. Whoever does not accept that women can equally speak for God (in public) is at odds with the world. The loss of external authority and gender roles fixed by divine authority inevitably leads to the loss of all truth. All attempts to impress male and female roles upon moderns seem oppressive, contrived, and irrational—because there is no divine authority in their minds.

"The Scriptures reveal timeless and eternal principles or doctrines of God that have authority over every generation, regardless of culture."[11] It is not that performing laws or works brings truth or salvation. The law (God's will) codifies God's inherent authority over man in a simple and direct way.

One modern scholar concludes a book on Scripture with this warning: "no rule governs [using Scripture], except the one against taking an individual passage in isolation from the biblical narrative."[12] When the overarching metanarrative deduced by the scholar overrules all the individual words, the scriptural authority is not allowed to overrule the authority of man and his rational religion. The issue is not whether God can speak, but whether man will hear. Christ says the problem of Scripture is not with Scripture at all:

> You will indeed hear but never understand, and you will indeed see but never perceive. For this people's heart has grown dull, and with their ears they can barely hear, and their eyes they have closed, lest they should see with their eyes and hear with their ears and understand with their heart and turn, and I would heal them (Mt. 13:14–15).

Nevertheless, the Word of God will bear still fruit, so that all is to Christ's glory (Lk. 8:15).

"The Gospel today is being used as a criterion for what is right and

---

[11] Tim Hegg, "The Role of Women in the Messianic Assembly" (TorahResource, 1988; http://www.torahresource.com/EnglishArticles/RoleofWomen.pdf), 9.

[12] R. W. Jenson, *On the Inspiration of Scripture*, 66.

wrong, what is true and false, what is permissible or not permissible."[13] The distinction between male and female, and their fixed duties, has become part of the discarded mythological framework of Scripture. Modern man who says nothing about God, in his hubris, will not allow any inflexible, mechanical authority over him. One scholar accepts female pastors by changing the wording of 1 Tim. 3:2 from: a pastor must be "the husband of one wife," to "having one spouse."[14] Another one positively states that "the truly orthodox Catholic Church, in its wisdom, has never included assent to complicated and personal ethical propositions in its creed."[15] The first Christian council did exactly that by appealing to the direct inspiration of the Spirit:

> For it has seemed good to the Holy Spirit and to us to lay on you no greater burden than these requirements: that you abstain from what has been sacrificed to idols, and from blood, and from what has been strangled, and from sexual immorality. If you keep yourselves from these, you will do well (Acts 15:28–29).

The rejection of male and female roles is a rejection of the God who "made me and all creatures." The evolutionary rationalization strengthens the philosophical case for a materialistic view of the world. The denial of authority, at its irrational and irreligious base, even allows the destructive criticism of equality to deny the body and natural law. The result is the acceptance of fluid gender divisions and the forced attempt to call male and female equal, when they obviously are not equal physically. Gender roles in sports and the military will remain controversial because women are not equal to men outwardly. Forced equality is decidedly un-egalitarian, but to speak the obvious is considered authoritarian, doctrinaire, and fundamentalist.

God's authority is what is being rejected by transgender and homosexual acts. St. Paul explicitly connects these sinful bodily acts to a "loss of the knowledge of God," that is, idolatry (Rom. 1). A theology which says nothing about our bodies or how we are to live authorizes practical atheism. To the educated modern following "consistent Cartesianism, God can only be known through an idea within man's inner, hidden

---

[13] A Concordia Seminary, St. Louis professor (1970–71), quoted in: P. A. Zimmerman, *A Seminary in Crisis*, 177.

[14] R. W. Jenson, *On the Inspiration of Scripture*, 56.

[15] Patrick M. Arnold in *The Fundamentalist Phenomenon,* ed. N. Cohen, 190.

ego."[16]

Absolute morality is foreign because Christianity has been billed as a non-authoritative philosophy, not the highest truth. God, though, does certainly speak to mundane things. Morality is not the Gospel, but false, ungodly morality can lead to a denial Christ. To continue in sin against the Spirit's direction is to re-crucify Christ as an unbeliever (Heb. 6:6). Belief, at its core, is submission to Christ's Word—and Scripture is His only present revelation. To believe what He says is to be subject to Him who gave His body in freedom. Without being subject to the Father of spirits there is no life (Heb. 12:14). One example of great faith was a man who understood authority from military service (Mt. 8:5–13). Christ compares the Gospel's authority to government and military authority! God's Word is God's authority in all things.

The false god of equality is based on the destruction of all authorities over man, especially God. To prevent God from speaking, the sanest procedure is modern atheism, which is essentially a denial of divine authority. Atheism is expressed today as a scientific judgment on God's existence, but to even entertain the question is to deny His actual authority as God. Historical criticism legislates that man can sit in the seat of God comfortably and decide such matters. However, unlike unreasonable theories of science, the assumptions are not discarded when the results are not consistent and do not make sense. Historical criticism is the shibboleth of modern idolatry because it legitimizes man's atheistic thoughts through the Cartesian method.

Yet the words of Scripture still stand. They can be ignored and denied, but they will not go away. It is not just the Gospel which remains, but the Word of the Lord. Male and female roles were divinely given at the very beginning of Creation, before the Fall into sin. "Those who dare to uphold the authority of Scripture . . . can expect today to be called bigoted, hateful, unenlightened, authoritarian, uncompassionate, and narrow-minded."[17] In declaring that women were made by Christ for a different earthly purpose than men (as a helpmeet and to propagate the human race), we uphold their equality in eternity, as well as God's direct action and authority in this world. Christ, who died a holy death,

---

[16] O. Valen-Sendstad, *The Word That Can Never Die*, 21.

[17] Stephen J. Hultgren in *Rightly Handling the Word of Truth,* ed. C. E. Braaten, 42.

did not make a mistake in creating males and females differently. The Spirit brings Christ equally, forgiving real sins against the Creator and the roles He specifically designed. The scriptural Christ does not wage war against external authority. He offers higher, heavenly comfort in the midst of misused authority.

The silence of women in public worship is mandated by 1 Cor. 14:34–35. The direct, doctrinal use of that passage is dismissed by Kloha: "Past use of the text within the LCMS has been in the propositional style exegesis, where the text presents divinely-inspired propositional statements devoid of historical setting, context or pragmatics." Of course, the critical scholar will attempt to bridge the historical gap—to speak holier words than the Holy Spirit. Scripture as merely an uninspired, human word will never say anything definite. "Scripture is not a legal text from which we extract value."[18] So the meaningless, "functional" definition of inspiration becomes quite the boon, because it can exist with any atheistic criticism. A consistent critic, who submits fully to his method, will reject not just Christ's words, but the Speaker of them.

It is not enough to rely on tradition, because tradition is authorized by sinners. We must prove our position from Scripture, not non-authoritative analogies. "Women preaching distorts the image of the incarnation."[19] This rational speculation is no doubt provocative, but it does not bind anyone or prove anything. Which "image of the incarnation" is the Christian bound to? The scriptural one.

The dissatisfaction with the text of the Bible is telling. One modern Lutheran use of Scripture, "common at the Fort Wayne [LCMS] Seminary," is the attempt to exclude "women from the teaching office on the basis of this kind of speculation [that the Son is subordinate to the Father] about the interior life of the Trinity as such."[20] If analogies rule, then perhaps pastors should be Jewish fishermen living near the Sea of Galilee. The incarnation and Trinity are not principles from which to derive all knowledge rationally. They are true facts defined and normed by Holy Scripture. As revealed truths, they can only be rejected or

---

[18] J. Kloha, "Text and Authority," 18.

[19] David P. Scaer, *Introduction to Method and Practice of Lutheran Theology*, 25.

[20] Mark C. Mattes, "A Review of *Let Christ be Christ: Theology, Ethics & World Religions in the Two Kingdoms. Essays in Honor of the Sixty-Fifth Birthday of Charles L. Manske*," *Journal of Lutheran Ethics* 2:6 (2002; http://www.elca.org/JLE/Articles/954).

believed.

Is it legalistic to say that God determines how we are to live in Christ? No. Legalism is man changing the Gospel, not God willing what is good and hating evil. Without fixed laws and divine, unbending authority, the Gospel turns to man-centered mush and Christ has nothing to forgive. The natural and universal gender roles are not the essence of Christianity, but neither can these be disregarded. Knowledge of Christ does not contradict God's revelation in our bodies. Our bodies are a divinely revealed book written by Christ. To deny God's authority over our existence is to lose the need and impetus for the Gospel. Only the Gospel revealed in Scripture saves and gives comfort to those who fail to act as God designed and still demands.

The fact that Scripture regulates behavior does not undo the Gospel. Without divine morality, the Gospel would be merely a mental construct. Man is not to prove inner coherence between revealed truths before accepting them. The Spirit and Son are one with the Father, so that the Spirit's words are Christ's. So also in worldly matters: the "authority of Scripture always remained the authority of its primary author."[21] The Father's authority over man, in all things, is part of the reason Christ submitted to the holy law for our sake. How could He save what He did not have authority over? Without external authority, grace is just an idea—not a reality.

Losing the male–female distinction leads to the loss of the Gospel. To lose authority, God-given roles, and specific directives based on biology is to lose the *need* for the Gospel. This is "an age which no longer asks, 'How can I get a gracious God?' "[22] But if we were to listen to God in Scripture and try to follow His strictures according to our created body, we would be desperate for grace. The impossibility of living up to the divine standard is humbling.

When doctrine is attacked, the real theologian must uphold the truth in face of destructive error. "Theology is no longer taken seriously," because man no longer needs God in the same way.[23] The only answer is to preach what is most hurtful, that man is not a god and his life is not his to live as he sees fit, otherwise he must face the wrath to

---

[21] M. Thompson, *A Sure Ground on Which to Stand*, 183.
[22] B. A. Gerrish, *Grace and Reason*, 99.
[23] K. Hagen, *Foundations of Theology in the Continental Reformation*, 228.

A Foundation for Confessing

come. Preaching must give people problems and what can only be seen as impossible demands—the imposition of divine authority over every aspect of their lives. Then the scriptural Gospel will be the answer to the problems of man.

Chapter 34

More Christian than Christ?

Modern theologians create systems. Each theologian criticizes his predecessor and makes a new rational gospel. The cycle goes on without end. However, the truth of God is given in a stable medium, so all updating of His truth is idolatry. "It is written" was enough for Christ (Mt. 4:4). Why do we demand more?

To be more Christological or Christocentric than the plain words of Scripture is to out-Christ the true Christ. It is to fashion an idol of what Christ should be to us and our reason. A modern idol is not made physically out of stone or wood but mentally, by arranging rational facts into a beautifully appealing system. We prefer ideas and themes we can twist and shape according to circumstances, not definite words that bind us. We desire to think our way to God by tailoring the truth to man. But like the idols of the Old Testament, the new idols, while appealing, are mute and deaf. "Christ becomes an imagined Christ, shaped by the religiosity and the unconscious desires of his worshipers."[1] An idol, by definition, does not speak: "You know that when you were pagans you were led astray to mute idols, however you were led" (1 Cor. 12:12). This is why muting Scripture is a basic premise of modernism: to leave only one god—critical man.

---

[1] James Smart, quoted in: R. A. Harrisville and W. Sundberg, *The Bible in Modern Culture*, 12.

Critiquing, not believing, is the essence of theology, according to reputable theologians. Authoritative truth has no place. A critical description of the content of the Scriptures, though, is no replacement for divine words. Without an origin in God, they are *not* His words or content. The founding certainty in theology comes from divine words, available in writing. Scripture norms all speaking about God, including the Gospel. We are "not at liberty to develop or deduce from this [material, Gospel] principle all other doctrines."[2] Without Scripture no doctrine has divine authority. What is not measured by Christ's inerrant Scripture is judged by man's supposed infallible critical authority.

Is Christ a person who died and rose and can therefore speak, or is this vocable merely a philosophical principle for speculation? Every modern craves the satisfaction of technical accuracy more than forgiveness. So we read the Bible critically and creatively, even if unintentionally, because we are bound in sin, alienated in our minds from God. However, Scripture was written to give life in Christ, not to appeal to the sinful flesh. So, "the evangelists did not purpose to recount everything in chronological order, neither did they do so in fact, . . . except where they have pledged themselves," as Luke did.[3]

Our ideas of how God should speak are irrelevant and inherently flawed, owing to our sin and hatred of God, who is the author of truth.

> Since, therefore, contempt of God, and doubt concerning the Word of God, and concerning the threats and promises, inhere in human nature, men truly sin, even when, without the Holy Ghost, they do virtuous works, because they do them with a wicked heart, according to Rom. 14:23: Whatsoever is not of faith is sin.[4]

Faith requires Christ's Word and Spirit. If Christ does not speak, what kind of savior could He possibly be?

The idolatrous danger is that the Cartesian method becomes the religion—a mere freeing from all authority. This secular gospel is license to think as God: "If the Christian gospel was to remain a living force in the modern world, it must be freed from dogma."[5] Doctrine, from whatever source, becomes the only evil influence. Christ the idea, but

---

[2] E. Hove, *Christian Doctrine* (Minneapolis: Augsburg, 1930), 29.
[3] J. M. Reu, *Luther and the Scriptures*, 49.
[4] AP IV:35.
[5] J. C. Livingston, *Modern Christian Thought*, 287.

not the factual person, is the philosophical lever to elevate man to godhood. "Idolaters . . . can make a god out of ideas as truly as the heathen make a god out of wood or stone."[6]

Repenting of speaking as a god, we must crucify reason and its thoughts and believe that "God does not speak to us except through the Scripture."[7] From this source the pure doctrine is to be taught, so that Lutherans "adorn the ministry of the Word with every kind of praise against fanatical men, who dream that the Holy Ghost is given not through the Word . . . ."[8] No matter what scholarly blessing it has, anything other than the external, physical Word of Christ is enthusiasm, and not of the true Christ. "For Jesus the Old Testament is the high court of authority to which all men could appeal," so that "even a word or a fragment of it was authoritative." This is strong proof for believing that Scripture is the mouth of Christ. Since Christ submits to it, it is not of less authority than the Lord Jesus Himself. After all, Jesus has "the tendency to identify God with Scripture."[9] The Book of Hebrews cites the Old Testament in proof-texting style by introducing a passage of Psalm 95 with "Therefore, as the Holy Spirit says" (3:7).

The modern itch to be more Christological than the true Christ is a deadly temptation. Christ has bound knowledge of Himself to words. Where an inner word takes precedence over the scriptural words, man finds himself worshiping a fictitious god, which exists only in the worshiper's wishful imagination. This is why we must voluntarily submit ourselves—not just to an idea about the Bible but to its actual words. They are Christ's words and by them we know Him and His grace. We do not subscribe to our own Christological fantasies, but to human confessions that are open for all to read.

What will we say, knowing that our confession will be judged on the Last Day? The Lutheran forefathers understood speaking God's Word as the most audacious and burdensome task:

> Since now, in the sight of God and of all Christendom, we wish to testify to those now living and those who shall come after us that this declaration herewith presented concerning all the controverted articles aforementioned and explained,

---

[6] C. E. Tulga, *Case against Modernism*, 24.
[7] R. D. Preus, *Inspiration of Scripture*, 4–5.
[8] AP XIII:13.
[9] A. A. Zaun, "A Study of the Idea of the Verbal Inspiration," 32, 43.

> and no other, is our faith, doctrine, and confession, in which we are also willing, by God's grace, to appear with intrepid hearts before the judgment-seat of Jesus Christ, and give an account of it.[10]

In the Spirit, who enlivens through the Gospel, we need have no fear of the modern world and its criticism. We have Christ's words and His righteous verdict spoken in our ears, therefore we have the same Christ who will receive us on the Last Day. This alone allows us to confess with confidence.

Obedience to Christ is obedience to what He says in Scripture. Scripture's words rule with all the authority of Christ, and He has given us no authority outside of Scripture. Luther warns:

> What punishment ought God to inflict upon such stupid and perverse people! Since we have abandoned his Scriptures, it is not surprising that he has abandoned us. . . . O Would to God that among Christians the pure gospel were known and that most speedily there would be neither use nor need for this work of mine. Then there would surely be hope that the Holy Scriptures too would come forth again in their worthiness.[11]

We do not need more statements merely referencing inerrancy and inspiration. Rather, like Justin Martyr: "We must not suppose that the language [of Scripture] proceeds from the men who are inspired, but from the Divine Word which moves them."[12] This teaching of inspiration leads us to take the divine words, and therefore Christ Himself, more seriously. "The church that doesn't love God's book commits suicide."[13] Luther explains why faith and Christianity are impossible without the divine authority of God's revealed Word:

> If you are not ready to believe that the Word is worth more than all you see or feel, then reason has blinded faith. . . . I do not feel the resurrection of Christ but the Word affirms it. I feel sin but the Word says that it is forgiven to those who believe. I see that Christians die like other men, but

---

[10] FC SD XII:40.

[11] Luther, *A Brief Instruction on What to Look for and Expect in the Gospels* (1521), LW 35:123–124.

[12] Quoted in: A. A. Zaun, "A Study of the Idea of the Verbal Inspiration," 76.

[13] William F. Beck, "Missouri's Decay" (1965), reprinted, *Christian News* 52:48 (Dec. 22, 2014), 5.

the Word tells me that they shall rise again.[14]

Without God's Word faith has nothing on which to rely.

"Little children, keep yourselves from idols" (1 Jn. 5:21). Humility is not scholarly, but it is Christian. Do not presume to tell God how He must speak to the educated, enlightened sinner. The Bible has a "simple and foolish guise, in order that He may quench all pride." We may castigate the Spirit or deny Him, but His humble words give sure knowledge of Christ.

> So the Holy Ghost has had to bear the blame of not being able to speak correctly but like a drunkard or a fool He jumbles the whole and uses wild, strange words and phrases. But it is our fault that we have not understood the language nor the style of the prophets.[15]

There are no more Christological words than those which the Spirit gave, unless Christ Himself be a liar: the Spirit of truth that speaks "will not speak on his own authority, but whatever he hears he will speak" (Jn. 16:13).

Submission to the divine Word allows for bold confessing, while "historical criticism produces only probable results. It relativizes everything."[16] Scripture, not as an object under our control but as God's own speech, changes the hearer. It very well could make him like Christ, hated by all the world. The Scripture's own method for understanding is quite unpleasant: "Affliction brings knowledge."[17] But the words of Spirit and life are not to be ignored. "I know no other Christ than the Christ of the Scriptures; and I know no other Scriptures than the Scriptures which Christ has given us."[18]

---

[14] Sermon on 1 Cor. 15:1 (March 31, 1529), quoted in: Adolf Köberle, *The Quest for Holiness: A Biblical, Historical and Systematic Investigation*, trans. John C. Mattes (Minneapolis: Augsburg, 1936; reprint, Eugene, OR: Wipf & Stock), 79.

[15] Luther, quoted in: J. M. Reu, *Luther and the Scriptures*, 29, 44.

[16] E. Krentz, *The Historical-Critical Method*, 67.

[17] M. Flacius, *How to Understand the Sacred Scriptures*, 69.

[18] E. F. Klug, "Holy Scripture," 127.

Chapter 35

Conclusion

"Why waste more words?" This thought occurred to me after seeing how much ink has been spilled on the doctrine of Scripture only to muddy the issue. However, since the truth of Christ has been given in the clear words of Scripture, it can be reformulated and applied in propositional fashion—and should be boldly confessed. The Holy Spirit is the power of Christ's teaching, not the human speaker or his mental powers. Heresy demands refutation, and if it is not rejected, the truth is quickly lost. This is not to say that true theology will ring true to sinners who hate the Author of truth, but when so many assume that God is a stuttering moron, incapable of facts or real truth, how can those who love God not speak out?

When God speaks, man must listen. "The [scriptural] revelation must be taken on its own terms, or not at all."[1] Christ's Word is truth, and He is known in words. Scripture—not the historical contingencies of its human writers—defines the extent and content of Lutheranism (pure Christian doctrine). Neither do our own times and concerns define or limit God's Word. Rather, the timeless " 'pure doctrine' is a legal title for the existence of the evangelical [Lutheran] church."[2] That the divine truth is expressed in specific historical, man-made writings is a Lutheran *fundamentum*. Scripture, inspired by the Spirit of truth, allows this possibility of confessing God's Word. A sinner may rationally speak of Christ, so that when the communication agrees with the divine Word

---

[1] Erwin Reisner, quoted in: C. E. Braaten, *History and Hermeneutics*, 37.
[2] H. Gensichen, *We Condemn*, 142.

it is God's Word and a sure vehicle for the Holy Spirit. The scientific study of the Bible allows no divine truth to emerge.[3] Historical criticism is unbridled atheism, since it is "insane to argue about God and the divine nature without the Word or any covering, as all the heretics are accustomed to do."[4]

"No recovery of Scripture is deep and genuine unless it entails also a recovery of the Confession" of what Scripture actually teaches.[5] The Lutheran Confessions do not derive their authority from Luther or their age, but through authentication by the Word of God. The Spirit alone works the conviction to acknowledge and confess the divine truth. "Scripture alone" does not mean that biblical expressions are the extent of our communication. Love of the truth revealed in tangible words explains the "Lutheran mania for making heretics," and also the necessity of confessing God's Word in present-day language.[6] Every Christian is to be a critic, not of God or His words, but for God and on the basis of His revelation.

The situation of modern times is outwardly different, yet essentially the same as in 1548, during the controversy of the Augsburg Interim. There are only two alternatives: "Peace or truth."[7] Since truth is of God, that means fighting with words to "destroy arguments and every lofty opinion raised against the knowledge of God, and take every thought captive to obey Christ" (2 Cor. 10:5). We need more polemics, begotten out of a love for the truth.

The Nicene Creed sufficiently explains inspiration: God spoke by His instruments, the prophets, then by the apostles. If "spoke by the prophets" contains an "indefiniteness of . . . expression," how can its other words, detailing the divine facts of the Christian faith, say anything definite?[8] When one phrase of the Nicene Creed is deemed unhelpful or incomprehensible, one must wonder how many other parts of that confession are being denied. Scripture's authorship was not under attack for the first 1,600 years of the Church, but the inspiration

---

[3] E. Linnemann, *Historical Criticism of the Bible*, 17.
[4] Luther, *Lectures on Genesis*, 1:13.
[5] Kurt Marquart in *Hermann Sasse: A Man for Our Times?*, ed. J. R. Stephenson, 186.
[6] H. Gensichen, *We Condemn*, 143.
[7] Letter written by Caspar Cruciger, quoted in: H. Gensichen, *We Condemn*, 145.
[8] H. P. Smith, *Inspiration and Inerrancy*, 100.

of Scripture was included as proof of the divinity of the Spirit: while "the Constantinopolitan Fathers did not at the time think that they were providing against a future evil, they were in truth supplying the Church with a weapon to be used in these last days, when the inspiration of the Holy Scriptures is being so ruthlessly attacked."[9] To deny inspiration is to deny one of the main proofs for the Spirit's divinity: "One of the most important roles played by the Holy Spirit in traditional pneumatology was the Spirit's inspiration of Holy Scripture."[10]

The word "inspiration," while it has received much lip service, "seemed more deference to traditional usage than a living sense of God's partnership in the process."[11] Since God is not allowed to interact with the world in modernist thinking, excluding the Spirit, both in the past inspiration of the biblical writers and the present enlightenment of the believer, is paramount today. Christ is allowed to remain an idea, a mental object, but where this "idol" touches and exercises authority over present man, there the satanic attack is felt. So Scripture, which for 1,600 years had been accepted as revelation, becomes a critical plaything for even the dullest of observers. To be Christian is either to critically fashion Christ with human words or to submit in humble reverence to the speaking Christ—and never the twain shall meet.

It may seem backward to deniers of inspiration, but those who defend Scripture have always used extra-biblical language to defend its teaching, because "controversies require more precise explanations."[12] It is a great irony that heretics want to stick to the simple language of Scripture to cloak their false teaching in religiosity. So extra-biblical words which destroy the arguments of heretics are not to be avoided, but cherished and confessed. It is not Scripture that requires them directly, but the pure understanding of God's Word that is upheld in the face of lies. Confessional words divide, because the real Christ divides. God's Word proclaimed here on earth will result ultimately in the division of the sheep and the goats. The words heard then will not be new to many critical goats, who will be sorely disappointed at the inerrantly speaking

---

[9] Alexander Penrose Forbes, *A Short Explanation of the Nicene Creed* (Oxford: John Henry Parker, 1852), 269–270.

[10] Abstract, W. L. Craig "Men Moved by the Holy Spirit Spoke from God," 1.

[11] This is spoken of the Torah. Eugene B. Borowitz in *The Fundamentalist Phenomenon,* ed. N. Cohen, 230.

[12] A. A. Zaun, "A Study of the Idea of the Verbal Inspiration," 71.

Christ. They will say: "Lord, when did we see you hungry or thirsty or a stranger or naked or sick or in prison, and did not minister to you" (Mt. 25:44). Note well that they will not say: "Lord, when did we hear you?"

There is either a multiplicity of limited, man-made theologies, or the singular, unlimited truth of Christ. "The mysteries of faith are objects of pure revelation and there is no faith without inner illumination."[13] When "faith is not [simply] assent to Scripture," but to the content revealed, the doctrine of inspiration must affect all theological conclusions and convictions.[14] Does the neglect of the doctrine of the inspiration of Scripture correspond to a deficit of the Spirit in modern Christians?

Scripture is weak and offensive according to man's reason. It is not given to fulfill every intellectual curiosity, but to lead to faith in Christ (Jn. 20:31). The desire to approach it as anything less than God's full and direct Word is a blunting of Christ Himself. The same Spirit of Christ continues to come to us in the reading and proclamation of His words. God's Word is not an abstract concept. The doctrine of inspiration will not lead into all truth, but the Spirit will, through His very own words. Only with a biblical doctrine of inspiration can Christians have confidence in what they believe. Faith is not in man's authority or wisdom, but in the power of the speaking God. We do not have to ascend to heaven to find God. He is here with us, right now, in His Word.

> God will not permit us to rely on anything else, to place our trust on anything which is not Christ in His Word, let it be ever so holy and full of spirit. . . . Only one thing will do, and that is that you turn away from yourself and all human comfort and throw yourself on the Word and nothing but the Word.[15]

The unity between Christ and Scripture is inherent: Luther "saw Scripture, [the] Word of God and God himself as one."[16] This is our Lutheran heritage, which we have sold for the porridge of academic respectability.

---

[13] Ludwig Crosius (1639), quoted in: Wim Janse, "Reformed Antisocinianism in Northern Germany: Ludwig Crocius' *Antisocinismus Contractus* of 1639," *Perichoresis* 3:1 (2005; reformatted: http://www.emanuel.ro/wp-content/uploads/2014/06/P-3.1-2005-Wim-Janse-Reformed-Antisocianism-in-Northern-Germany.pdf), 8.

[14] H. D. McDonald, *Theories of Revelation*, 2:343.

[15] T. Engelder, "Holy Scripture or Christ?," Part 1, 493.

[16] K. Scholder, *Birth of Modern Critical Theology*, 21.

"Since it is God's Word, it must be regarded as no less high and venerable than God Himself."[17]

Historical-critical methodology dissolves the truth of theology. Perhaps those who are inconsistent can rely on the weak crutch of tradition, but over generations the effects will be devastating. Only when Christians are convinced of the truth will ministers authoritatively proclaim it, so the Church is properly rooted in Christ alone.

The final result of historical-critical principles, the belief in "the inadequacy and brokenness of all human attempts to state the truth," disarms theology.[18] It leaves man with nothing to say and no Gospel. "Righteousness is wrought in the heart when the Holy Ghost is received through the Word."[19] Where there is no physical Word of Christ, the living Christ is allowed no interaction with living people. This is the paramount heresy of our time. But "there must be heresies if the Church is to clarify her teachings."[20]

Truth is on the side of the persecuted, not the persecuting—so bravely endure the charges of "fundamentalist" and "anti-intellectual." Jesus Himself was more of a fundamentalist than anyone alive today. Yet we confess Him Lord of earth, truth, and our very bodies—won by factual, divine blood. He who has ears to hear, let him hear. He who has a tongue to speak, let him speak according to what the Spirit said and moves us to repeat: Christ died for sinners who are against Him, and calls all to un-modern repentance, which has not changed since Adam hid from the Lord in mind and body.

The teaching of Luther and the Confessions preaches against modern atheism. Their confession leads into God's Word, and can show the remnant how to take the Word seriously. If the price of confessing the truth is being called "medieval," "biblicistic," or "a fundamentalist," so be it. "Blessed are you when others revile you and persecute you and utter all kinds of evil against you falsely on my account. Rejoice and be glad, for your reward is great in heaven, for so they persecuted the prophets who were before you" (Mt. 5:11–12). This is a word to rely on, or to be trampled by—there is no middle ground in this spiritual

---

[17] Luther, Commentary on Ps. 111:10, LW 13:362.
[18] David L. Tiede in *Studies in Lutheran Hermeneutics,* ed. J. Reumann, 283.
[19] AC XVIII:3–4.
[20] B. A. Gerrish in *The Seven-headed Luther,* ed. P. N. Brooks, 17.

battle. The truth is not an abstraction. It is life that brings Christ to the believer and makes the sinner new before God. All atheistic thoughts are to be crucified with Christ. God's Word fills His children with hope in the Christ-centered world to come.

Idolatry is not less common today than in any other age, but today idols are mentally formed. This deformation of the spiritual Gospel is the same man-made idolatry condemned as futile in the Scriptures: "Their idols are silver and gold, the work of human hands. They have mouths, but do not speak; eyes, but do not see" (Ps. 115:4–5). "Christ" as a wordless idea is the modern idol that must be destroyed with words of true power and authority. The Spirit will work when and where He wills through this Word.

We must speak God's words and be prepared to die with Christ, lest we re-crucify Christ by critically crucifying His words over and over again. There is no safe ground between God's Word and man's word. "We owe to the Scripture the same reverence which we owe to God, because it has proceeded from him alone, and has nothing of man mixed with it."[21] In surveying the morass of modern teaching on Scripture, one may rightly summarize the modern crucifixion of the Scriptures: "Ah, one sees, Satan is making his final efforts to take away from Christians their full treasure, for he does not have much time."[22]

Inspiration is not the most important truth, but it is part of God's truth, and one that cannot be taken for granted in these last days. The Reformation clarified the doctrine of justification, following in the footsteps of the Early Church's Christological and Trinitarian confessions. The Church of our day must wrestle with the Spirit's role in history, especially His dictating of the words of Scripture to passive, yet cooperating, men. True "sacred theology, though it is doctrine divinely inspired, does not shrink from being treated in letters and words."[23] Not only is Christ incarnate, but also His truth, "which we have heard, which we have seen with our eyes, which we have looked upon, and our hands have handled" in Scripture (1 Jn. 1:1). When inspiration is denied, the doctrine of the Spirit's work in delivering Christ's righteousness and enlightening the believer will be deformed.

---

[21] John Calvin, quoted in: *Inerrancy and the Church,* ed. J. D. Hannah, 161.
[22] C. F. W. Walther, "Evening Lectures on Inspiration," Lecture VI.
[23] Luther, quoted in: H. Gensichen, *We Condemn,* 52.

Since we have God's own Word, let us speak it with unabashed confidence. That is the task for true, scriptural theologians, even if it means criticizing the most brilliant critics. Scripture is God's mouth for us, therefore let us be bold to speak the truth.

> For if they believed that these [words of Scripture] were God's words they would not call them 'poor, miserable words,' but would prize a single tittle and letter more highly than the whole world, and we would fear and tremble before them as before God himself. For he who despises a single word of God certainly prizes none at all.[24]

Because God speaks clearly, we, at the Spirit's urging, do too. Amen.

All flesh is like grass
and all its glory like the flower of grass.
The grass withers, and the flower falls,
but the word of the Lord remains forever.
And this word is the good news that was preached to you.
1 Pet. 1:24–25

---

[24]Luther, *Confession Concerning Christ's Supper* (1528), LW 37:308.

Acknowledgments

Many people have helped and encouraged me on this journey of not only writing, but continuing to stand in the pastoral office. First, I must express my deepest gratitude to my wife Aubri for supporting and tolerating my absence during the writing process, and for being a faithful wife and mother in submission to Christ. Fellow pastors and their encouragement to write have been invaluable, especially that of Clint Poppe, Herman Otten, and Steven Flo. In seminary professor Roland Ziegler fostered the exploration of this topic academically. Thanks are due to Brent Kuhlman, who tasked me with presenting on this topic for the Nebraska Lutheran Confessional Series, and to the attendees who have been willing to listen to me. I am grateful for the saints at Zion Lutheran Church, Omaha and Lance Berndt, a truly scriptural theologian, who have been patient with their new associate pastor and have allowed me the time to read and write. My editors Peter Stafford and David Waller were professional and diligent in their work. I am very thankful for the volunteer readers of this and of my previous writings, whose feedback has been invaluable, especially Clint Stark, Daniel Lepley, and Daniel Speckhard. Weslie Odom graciously designed the cover and urged me to press on with this work. Lastly, thanks are owed to my parents, Neil and Verna, and to all the faithful pastors who have followed them in patiently teaching me the Word of Christ.

Appendices

Index of Subjects

anthropology, 104, 112, 115, 116, *see also* Pelagianism
    and inspiration, 189
    based on the Fall, 119
    gender roles, 278, 280, 283
    not capable of God's Word, 71
    sin, 67, 139
apologetics
    avoided by moderns, 243
atheism, 81, *84*
    modern, 281
    practical, 196, 275, *280*
authority, 21, 151
    in words, 67, 228
    of Christ, 109, 142, 155, 210
    of the Gospel, 262, 277
    parental analogy, 207, 244
    submission to, 277
certainty, 252
    a spiritual problem, 92, 96, 143
    basis of, 288
    Cartesian doubt, 83
Christianity
    is factual, 81
    absoluteness of, 76
    role of history, 73, 119
    the Bible-believer, 61, 97
Christology, 206
    Christ, *228*, *231*, *235*
        a philosophical principle, 169, 201, 202, 229, 286
        known through words, 162
    Christ the Word, 176, 203
    incarnation, 105, 113, 235
    path to, 249
    resurrection, 236
Church, 189, *193*
churches
    doctrinal differences, 92
    fellowship, 192
    LCMS, iv, 146, 223, 226, 229, 236–239, 250, 253–255, 257, 270, 282
        inspiration in, 225
        neo-orthodox language in, 234
        programs in, 237, 253
        Seminex, *206*, 229, 238
    Lutheran, *261*, 269
    Reformed
        communication of attributes, 96
        doctrine of Word, 96
        Lord's Supper, 96, 169
Communion, *see* sacraments
confessing, 170, *233*, *261*
    has been silenced, 269
    in view of the Last Day, 271, 287, 293
    is intolerant of pluralism, 80
    is necessary, 293
    requires condemnation, 257
    requires the Spirit, 157, 288
    with extra-biblical words, 293
confessions, *232*, *270*
    Chicago Statement, 148
    Lutheran, 98, 233, *267*
        Apology of the Augsburg Confession, 134, 175, 182, 186,

232, 286
Augsburg Confession, 118, 136, 169, 234, 287, 295
critical translation, 232
Formula of Concord, 5, 113, 142, 175, 186, 193, 208, 222, 232–234, 265, 272, 288
Saxon 1549, 43, 46
Smalcald Articles, 224
Small Catechism, 131, 168, 181, 227, 247
Nicene Creed, iv, 3, 10, 44, 52, 108, 230, 233, 254, 292
purpose of, 273
quia subscription, 261, 269, 270
doctrine, *161*, 180, 273
basis of, 11, 173
is revealed, 231
unity of, 265
ecumenism, 192, 270
education, 78, 119, 137
theological, 184, 225
Enlightenment, the, 68, 69, 104, 107, 115, 221, 235
equality, *see* anthropology
exegetics, *see also* interpretation
scholarly, 63
commentaries, 173
sweat is seed of heresy, 249
themes, 245
sacramental, 238, 245
facts, 36, 251
verification of, 67, 79, 80
faith, 95, 154, *168*, 183, 205, 252
depends on the Word, 110
is divine certainty, 264
is problematic for moderns, 233
is submission to authority, 262
scientific, 84
freedom, *see also* Gospel, 160
only for man, 112
fundamentalism, *12*, 116, 262, 272
causes division, 109
has absolute authority, 208
historical definition, 26
is orthodoxy, 262
is pre-modernism, 33

persecution, view of, 170
used as insult, 87, 295
God
communicating, *91*, 100, 106, 142
known only in revelation, 94
monergism, 113
only a hypothesis, 104
Spirit's divinity, 293
Trinity, 109
Gospel, 188, 195, *207*, 211, 239, *277*, 283
and Scripture, 161
authority of, 278
freedom of, 32
is propositional, 206
is submission to authority, 95
only one, 206
redefined, 113, 197
hermeneutics, *129*, 132, *136*, *see also* exegetics
works righteousness of, 133
historical criticism, 75, *84*, 85, 252, 281
a method of universal doubt, 84
all or nothing, 85
allows no authority, 68, 166
allows no confessing, 289
allows no revelation, 85
allows only human sources, 83
assumed anthropology, 116, 122
atheistic neutrality of, 82
based on doubt, 263
creates multiple theologies, 245
divine–human split, 201, 214, 218
identified with the Gospel, 186
in Islam, 122
is undogmatic, 248
its definition of history, 66
makes man equal to God, 246
prevents confessing, 269
redaction criticism, 229, 244, 246, 248
sacred status, 87
history
historical consciousness, 67

304

scientific, 59, 65, 66, 85, 169
idolatry, 296
    hermeneutic, 245
    mental, 65, 114, 133, 180, 287
        the idea of Christ, 296
    modern, 31, 84, 202
    of method, 87
    rational, 84
    sacramental, 187
    scriptural, 62
inerrancy, 11, 21, 23, *72*, 93, 106, 143, 264
    a confession of faith, 169, 212
    a property of communication, 13, 15
    a property of truth, 23
    associated with reason, 82
    false origins of, 9, 44
    not a scientific claim, 79
    not Christian, 15
    of Christ, 26, 161
    of doctrine, 15
    of man, 36
    relation to inspiration, 15
inspiration, *26*, *44*, 50, *220*
    analogous to the Supper, 25
    and authority, 96
    and science, 13
    autographs, 145, 148, 152, 226
    conversion analogy, 112, 222
    dictation, 42, 51
    dual authorship, 108
    false alternatives to, 155
    functional, 282
    impacts all doctrine, 230
    in Lutheran Confessions, 175
    is foundational, 137
    means revelation, 45, 107
    musical analogy, 51
    not mechanical, 103, 104, *112*, 220
    offense of, 43
    paid lip service, 293
    proof of Spirit's divinity, 292
    rules out criticism, 123
    the command to write, 110
    verbal, 45, 106

interpretation, *see also* exegetics, *171*
    literal sense, 167, 171
    not a rational method, 141
    proof-texts, 168, 176, 279
    requires the Spirit, 97, 131
    role of affliction, 162, 176
    scriptural method, 289
Islam, 3, 118
knowledge, basis of, 58
language, 104, 145
law
    and Gospel, 186
    third use of, 185
liturgical movement, 181, 183
liturgical theology, 182
mathematics, 108, 109
method, Cartesian, 85
ministry, office of, 190, 282
    ordination, 191, 275, 278, 282
    preaching, 196, 228, 295
miracle, *31*, *69*, 70
    defines God, 86
modernism, 12, *33*, 47, 48, 60, 104
    is enlightenmentism, 30
    bedrock assumption of, 72
    distinction of fact and value, 82
    how it persecutes, 214
    is anarchist, 277
    is bad anthropology, 104, 115
    is fanaticism, 92
    is mathematical, 108
    is Pelagianism, 117
    is super-Arminianism, 103
    knowledge redefined in, 31
    mindset of, 62
    post-modernism, 30, 129, 145, 249, *250*
    right to judge, 61
    separation of Word and Spirit, 96
    synthesis of all heresies, 10, 23
morality, 206, 266, 275, 280, *283*
    abortion, 6
    sexual, 163
Muratorian Canon, 247
neo-orthodoxy, *25*, 169, 208, 234, 235, 251, 266

no truth in, 264
revelation in, 271
oracles, 10, 31, 35, 53, 58, 139, 167, 173, 226, 238
in philosophy, 110
Pascendi Dominici Gregis, 10
Pelagianism, *116*
super-Pelagianism, 117, 118
persecution, 268, 289, 295
role of division, 80
philosophy, 12
finite not capable of infinite, 234
pneumatology, 88, 133
illumination, 115
in Christian life, 227, 294
testimony of the Spirit, 87, 88, *143*, 157
polemics, *225*, 273
post-modernism, *see* modernism
programs, *253*
prolegomena, 9
propositions, *29*, 161, *228*, 250, 261, 266, 271
Providentissimus Deus, 10
rationalism, *27*, 180
reason, 216
enemy of theology, 220
replaces Holy Spirit, 243
secondary *principium* for Reformed, 28
the modern god, 117
worship of, 27
repentance, 241
revelation, 12, 37, 73, 271
and reason, 41
criteria of, 69
denial of, 119
general, 69, 71
is the final arbiter, 166
natural law, 276, 277
personal, 237
propositional quality of, 23
role in theology, 42
self-revelation, 93, 105, 111
special, 59, 116
two books analogy, 70
unity of general and special, 79

unverifiable, 58, 61
right of private judgment, 58, 97, 119, 167
sacraments, 187, 233, 238
Communion, closed, 232, 257
Lord's Supper, 172, 183, 232, 268
science, 12, 35, *57*, *see also* history, *59*, 63, 97, *109*
attitude towards objects it studies, 95, 109
cosmology, 78
effect on theology, 57, 61
evolution, 81, 119, 280
limitations of, 67
method of, 22, 59, *83*
natural laws, 31, 69, 118
philosophical basis of, 109
religion of, 77, 78
scientism, 59, 256
scientists, 22
natural, 48
religious beliefs, 22
Scripture, 209
canon, 152, *154*, 156
antilegomena, 151, 156
homologoumena, 153
Christological analogy, 117, 213–215, 219, 223
clarity of, 98, 140, 162, 195
dynamic and static properties, 24–26
function of, 161, 210
has been secularized, 119
is God's mouth, 163, 175
is its own interpreter, 171
is timeless, 96
"it is written", 99, 124, 167, 205, 267, 285
Jesus' use of, 43, 143, 148, 185, 197, 207, 210, 241, 285, 287
not docetic, *112*, 216
passages differ, 171
role of difficulties, 171
science and history, 25, 161
sola Scriptura, 138, 175, *183*, 239, 248
textual criticism, 147

touchstone, 175
translation, 148
unity of, 93, 98, 158, 248, 286
writers as instruments, 51
sin, *see* anthropology
Socinianism, *48*, 115, 159
  is modernism, 204
sola Scriptura, *170*
Spirit, Holy, *see* pneumatology
technology, 62, 67
theology, *23*, *161*, *166*
  authority of, 23, 88
  basis of, 141
  Christ not the norm of, 169
  is factual, 26, 251
  is propositional, 28
  is revealed, 163, 240
  is timeless, 36
  not a system, 98, 162
  not academic, 240
  not creative, 123
  not scholarly, 217
  queen of the sciences, 46
  requires revelation, 51
  requires the Spirit, 136
  revealed, 94, 163
  role of reason, 170
  Scripture the *principium*, 170
theology, modern, 32, 36, 65
  assumes multiple theologies, 155, 197
  false principles of, 185, 187
  hero worship, 255
  is a secular subject, 94
  is academic, 119
  is creative, 91, 256
  is systematic, 184, 202
  language of, 245
  narrative theology, 249
  only speaks of man, 94
  scholarship, 196, 238
tradition, 42, 52, 94, 173, 257, *282*, 295
truth, 13, 24, 25, 72, 160, 211, 264
  incarnate, 296
  non-factual, 82
  propositional, 266
  rational, 83
  redefined, 80, 82
vocation, 188
Word, *98*, 138, 287
  and Spirit, 131
  basis for the Church, 158
  basis of, 96
  false criteria, 155
  is God to man, 83
  not accepted, 98
  pastoral authority, 185
worship, 253
  highest, 134

Index of People

Adams, David L., 74, 131
Ambrose of Milan, 221
Aquinas, Thomas, 50, 110, 217
Arnoldus, Nicolaus, 117, 131, 189
Augustine, 16
    clarity of Scripture, 171
    history, 66
Bacon, Francis, 36, 81, 85
Barth, Karl, 25, 75, 82, 192, 201, 202, 231, 234–236, 243, 253
Basil of Caesarea, 35
Bavinck, Herman, 141
Bettex Frédéric, 27, 36, 37, 43, 71, 78, 97, 104, 142, 241
Bombaro, John, 209, 210, 239, 255
Bonhoeffer, Dietrich, 243, 253, 276
Brunner, Emil
    inspiration is greatest tragedy, 79
    Scripture, two natures, 219
    evolution is truth, 78
    history defined in Jesus, 202
    mass atheism, 32
    revelation only event, 73
    revelations for pre-moderns, 31
    Scripture idolized, 105
    theology is secular, 94
    verbal inspiration felled by science, 91
Brunner, Peter
    Church is silent, 274
    no confessional subscription without Scripture, 256, 270
Bultmann, Rudolf, 94, 192, 229, 235

Calov, Abraham, 204
Calvin, John, 49, 156, 296
Chemnitz, Martin, 49, 50
Clement of Rome, 51
Coleridge, Samuel Taylor, 71, 75, 202
Crosius, Ludwig, 294
Cruciger, Caspar, 292
Cyrus, King, 74
Dannhauer, Johann Conrad
    norm for Christ, 204
Descartes, René, 59, 62
Dilthey, Wilhelm, 48, 132, 159
Donne, John, 167
Ebeling, Gerhard, 60, 93, 122, 165, 180, 253, 254
    confessing, 263, 270, 272
    everything is historical, 74
    hermeneutics, 196
    historical critical method, 68, 79
    resurrection, 256
    revelation, 73
        is docetic, 186
    truth, 76
Engelder, Theodore, 17, 25, 27, 103, 110, 116, 124, 168, 184, 197, 204, 207, 220, 226, 267, 294
Eugene IV, Pope, 42
Flacius, Matthias, 138, 161, 163, 171, 173, 289
Fosdick, Harry Emerson, 11, 17
Franzmann, Martin, 223
Fuller, Andrew, 105
Gadamer, Hans-Georg, 83
Gerhard, Johann, 100, 106, 161, 163

Gibbs, Jeffery, 250
Gieschen, Charles, 156
Girgensohn, Karl, 82
Green, Lowell, 247
Gregory the Great
   canon, 154
   Scripture like river, 144
   Spirit is author, 44
Hardt, Tom G. A., 36, 217
Henry, Matthew, 43
Hermann, Wilhelm, 168
Hippolytus of Rome, 51
Hodge, Charles, 180
Hugh of St. Victor, 70, 138
Hunnius, Aegidius, 48
Ignatius of Antioch, 267
   gospels are flesh of Christ, 209
Irenaeus, 53
James Smart, 285
Jenson, Robert W., 190, 279
Jordahl, Leigh D., 5, 6, 236
Justin Martyr, 169, 221, 288
Kepler, Johann, 78
Kierkegaard, Sören, 173
Klein, Ralph, 12, 276
Kloha, Jeffery, 118, 146, 153, 205, 223, 226, 227, 237
   authority is legalistic, 227
   authority of Scripture, 205
   historical criticism, 282
   secondary divine authority, 151
Klug, Eugene F., 45, 88, 174, 175, 215, 240, 270, 289
La Peyrère, Isaac, 116
Lake, Kirsopp, 16
Le Clerk, Jean, 117
Leo XIII, Pope, 46, 173
Lessing, Gotthold Ephraim, 73
Lewis, C. S., 251
Locke, John, 61
Löhe, Wilhelm, 191, 254
Luther, Martin, 227
   a medieval theologian, 42, 47, 144
   age of world, 47
   assertions, 96
   be satisfied with text, 172

Bondage of the Will, 139
   cross necessary, 176
   crucifying text, 231
   do not separate God and man, 107
   faith in Word, 288
   follow God's command, 172
   implicit trust in the Word, 99
   inspiration, 44, 139, 221
     in Nicene Creed, 52
   martyrdom for written words, 195
   on James, 155
   propositions, 271
   reason invents a god, 218
   research of, 44, 181, 267
   revelation, 53
   role of reason, 27, 170
   Scripture, 219, 286, 294
     clarity of, 176
     find Spirit in, 97, 201
     gives confidence, 189
     God punishes its despisers, 288
     harmonization of, 248
     impossible to man, 174
     interpretation of, 172, 174, 273
     is a worm, 142
     is Christ to me, 179
     is God's mouth, 297
     proof-texts, 185
     Satan causes confusion, 256
     sola Scriptura, 188, 294
     sources, 222
     textual criticism, 147
     unity of, 99, 183
     value of letters, 144
     word clothing of Christ, 209
     words of, 156, 173
   signs prove revelation, 210
   source of all heresy, 92
   Spirit
     is interpreter, 97
     is literary author, 52, 131, 240, 289
     is necessary, 137

wiser than man, 41
Supputatio annorum, 47
theologians made by letters, 161
theology, 46
   in letters, 296
   is scriptural, 241
   theology of, 267
   Word and Spirit, 224
   Word received by unbelievers, 264
Machen, J. Gresham, 81, 219
Maier, Gerhard, 45, 116, 117, 147, 195, 206, 211, 214, 218, 245
Major, Johann, 138
Marcion, 228
Marquart, Kurt, 71, 86, 123, 216, 266, 269, 292
Melanchthon, Philip, 48, 232
   reason, 220
Micraelius, Johannes, 116
Miller, John
   interpretation is not Scripture, 196
   revelation, 119
   spiritual idolatry, 32
   submission to Gospel, 133
Milton, John, 50
Nafzger, Peter, 24, 44, 72, 74, 129, 155, 157, 188, 206, 208, 211, 229, 231, 234, 236
   Bible not the Word of God, 251
   Christ is revelation, 252
   God undefinable, 234
   no divine words, 104
   on Luther, 47
   Scripture only a functional authority, 24
   search for third way, 251
   theology undeveloped, 240
Niebuhr, Reinhold, 72
Noland, Martin R., 98, 219, 270
Oepke, Albrecht, 130, 143
Okamoto, Joel, 236
Outler, Albert, 93
Pfeiffer, August, 174
Pieper, Franz, 236
   Church, 158
   faith, 110
   inerrancy, 161, 169
   instruments of Spirit, 51
   insulted, 5, 237
   judgment on critics, 196
   Lutheran Confessions, 271
   reason, 162
   Scripture, 6
   sola Scriptura, 173
   textual criticism, 145
Piepkorn, Arthur Carl, 235
Piper, John, 139
Plato, 110
Preus, Robert, 52, 53, 111, 152
Quenstedt, Johannes Andreas, 157
Reisner, Erwin, 68, 82, 85, 265, 291
Ross, Alexander, 60
Sagan, Carl, 57, *58*, 59, 256
Sandeen, Ernest R., 21
Sasse, Hermann, 118, 162, 215–218, 223, 255
   Aquinas, 217
   Christological analogy, 215
   fundamentalism, 216
   inerrancy controversy, 216
   scientific assumptions, 217
   Scripture, 215
   search for third way, 215
Saunders, Brian, 225
Scaer, David, 88, 240, 245
   criticism, 229
   exegesis without norm, 245
   Gospel is standard, 238, 241
   history, scientific, 229, 238
   inspiration not direct, 227, 228
   John 6, 244
   misuse of Luther, 44, 241
   multiple words of God, 221
   ordination, 191, 282
   proof of Scripture's character, 204
   redaction criticism, 246, 248
   testimony of the Spirit, 87, 88
   theology from below, 238
   theology is systematic, 184
Scaer, Peter, 247
   harmonization, 232

Scharlemann, Martin, 112
Seeberg, Reinhold, 124
Semler, Johann Salomo, 166
Seraphim, Rose, 24
    theology v. science, 12, 24
Smith, Henry Preserved, 11, 95, 104, 123, 292
Socinus, Faustus, 48
Spinoza, Baruch, 30, 221
St. Bernard of Clairvaux, 206, 212
    interpretation, 137
Thielecke, Helmut, 78, 95, 124, 184, 236, 264, 268, 278
Tillich, Paul, 69
Turtullian, 249
Valen-Sendstad, Olav, 75, 135, 281
Voelz, James, 139, 146, 219, 223, 227
    confessing, 233
Voetius, Gisbertus, 28
Walther, C. F. W., 142, 172, 187, 190, 191, 296
    Alleosis, 223
    certainty, 143
    divine–human Scripture, 223
    fate of LCMS, 257
    historical criticism, 122
    inerrancy, 262
    inspiration, 10
    Lutheran Confessions, 267
    nuda Scriptura, 181
    Scripture, 271
Wolff, Hans Walter, 88
Zwingli, Ulrich, 218
    Alleosis, 223

Index of Bible Verses

Genesis
   1:28, 28
   3:5, 277
   18:25, 228
   30:2, 240
   49:11–12, 139
Exodus
   4:11–12, 110
   17:14, 110
   34:27, 110
Numbers
   21:14, 222
Deuteronomy
   8:3, 246
Joshua
   10:12–14, 48
   10:13, 222
1 Kings
   5:19, 147
2 Chronicles
   36:22, 74
Job, 44
   12:22, 218
   32:8, 218
Psalms
   34:1–22, 51
   51:4, 113
   56:1–4, v
   90:1–17, 52
   95:7–11, 287
   111:10, 295
   115:4–5, 296
   119:71-72, 175
Isaiah
   8:1, 110
   9:2, 47
   30:8, 110
   53:5, 245
   55:11, 230
Jeremiah
   30:1–2, 110
Matthew
   2:23, 205
   4:4, 285
   5:8, 119
   5:11–12, 295
   5:13, 258
   5:18, 143
   7:21, 252
   8:5–13, 281
   8:8–10, 100
   10:29, 74
   10:32, 261
   11:30, 225
   13:14–15, 279
   16:23, 180
   19:3–5, 148
   19:4, 277
   19:26, 105, 171
   21:16, 138
   22:43, 113
   23:15, 236
   23:34–39, 241
   24:14, 206
   25:31–46, 271
   25:44, 294
   26:53–54, 205
Mark
   8:38, 141
   12:27, 237
   12:42, 142
   14:21, 205
   16:16, 247
Luke
   1:70, 3
   4:3–13, 43
   8:12, 225
   8:15, 279
   16:29, 271
   24:32, 45, 211
John
   1:1, 176
   3:8, 133, 136
   4:23, 181, 183
   6:1–71, 233, 244
   6:45–46, 136
   8:31, 265
   8:31–32, 168
   14:6, 163
   16:12, 143
   16:13, 289
   16:14, 205
   20:24–29, 61
   20:31, 294
Acts
   4:31, 157
   13:20, 147
   15:28–29, 280
   17:28, 69
Romans
   1:18–32, 280
   1:20, 70

1:21–23, 195
2:16, 267
3:2, 31
7:12, 185
8:11, 227
8:21, 276
10:8, 230
10:8–10, 269
10:17, 267
13:1, 52, 277
14:23, 286
15:4, 83
16:25, 267
16:25–26, 95
1 Corinthians
  1:20, 132
  2:11, 113
  2:12, 113
  2:12–13, 137, 271
  2:15, 189
  6:9–11, 278
  11:3, 276
  11:19, 80
  11:29, 232, 233
  12:3, 113, 157
  12:12, 285
  13:12, 161
  14:34–35, 282
  15:1, 289
  15:12–19, 236, 251
  16:21, 220

2 Corinthians
  2:17, 268
  3:17, 112
  10:5, 226, 292
  12:4, 131
Galatians
  1:6–7, 208
  1:6–9, 265
  2:20, 267
  4:6, 268
  5:1, 211
  5:9, 247
  5:19–21, 278
  5:21, 160
Ephesians
  2:5, 116
  3:12, 263
  5:25–27, 189
Philippians
  2:10, 160
  3:8, 136
Colossians
  1:16–17, 86
1 Thessalonians
  2:13, 188
  4:8, 163
1 Timothy
  2:12, 278
  3:2, 280
  4:1–7, 172
2 Timothy
  2:8, 267

3:7, 89
3:16, 158, 160, 173
4:3, 255
Hebrews, 158
  1:1, 50
  1:1–2, 197
  1:1–3, 32
  1:3, 69, 71
  2:6, 100
  3:7, 287
  6:6, 281
  12:14, 281
  12:29, 94
  13:7–9, 263
James, 155
1 Peter
  1:24–25, 297
  2:24, 138
2 Peter
  1:21, 107, 111, 154, 158, 224
1 John
  1:1, 296
  5:7–8, 147
  5:21, 289
Revelation
  22:18–19, 157, 174

References

Abraham, William J. *Divine Revelation and the Limits of Historical Criticism.* New York: Oxford University Press, 1982.

Allen, Diogenes. *Philosophy for Understanding Theology.* Atlanta: John Knox Press, 1985.

American Heritage Dictionary of the English Language. 5th ed. http://www.thefreedictionary.com/propositionally.

Angier, Natalie. *My God Problem.* http://edge.org/conversation/my-god-problem. Nov. 19th, 2006.

Aquinas, Thomas. *The Summa Theologica of St. Thomas Aquinas.* 2nd rev. ed., 1920. http://www.newadvent.org/summa/3171.htm.

Avis, Paul, ed. *The Science of Theology.* Vol. 1 of *The History of Christian Theology.* Grand Rapids, Eerdmans, 1986.

Baier, Johann Wilhelm. "A Compendium of Positive Theology." Edited by C. F. W. Walther. Translated by Ted Mayes. Unpublished draft, 2012.

Baillie, John. *The Idea of Revelation in Recent Thought.* New York: Columbia University Press, 1956.

Barr, James. *Beyond Fundamentalism: Biblical Foundations for Evangelical Christianity.* Philadelphia: Westminster Press, 1984.

Basil of Caesarea. "Hexaemeron Homily." In *Nicene and Post-Nicene Fathers.* Second series. Vol. 8. Edited by Philip Schaff. Buffalo, NY: Christian Literature Publishing, 1895. http://www.newadvent.org/fathers/32019.htm.

Becker, Siegbert W. *The Foolishness of God: The Place of Reason in the Theology of Martin Luther.* Edited by John A. Trapp. 2nd ed. Milwaukee: Northwestern Publishing House, 1997.

———. *The Scriptures—Inspired of God.* Milwaukee: Northwestern Publishing House, 1971.

———. "The Word of God in the Theology of Martin Luther." Unpublished paper, 1963. http://www.wlsessays.net/bitstream/handle/123456789/363/Becker Theology.pdf.

Behrens, Achim and Jorg Christian Salzmann, eds. *Listening to the Word of God: Exegetical Approaches*. Göttingen: Edition Ruprecht, 2016.

Beiser, Frederick C. *The Sovereignty of Reason: The Defense of Rationality in the Early English Enlightenment*. Princeton: Princeton University Press, 1996.

Belt, Henk Van Den. *The Authority of Scripture in Reformed Theology: Truth and Trust*. Leiden: Brill, 2008.

Benoit, Pierre. *Aspects of Biblical Inspiration*. Translated by J. Murphy O'Conner. Chicago: Priory Press, 1965.

Bente, F. and W. H. T. Dau, eds. *Concordia Triglotta: The Symbolical Books of the Evangelical Lutheran Church*. St. Louis: Concordia Publishing House, 1921.

Berkhof, Hendrikus. *Two Hundred Years of Theology: Report of a Personal Journey*. Translated by John Vriend. Grand Rapids: Eerdmans, 1989.

Bettex, Frédéric. *The Bible the Word of God*. Translated from 3rd. German ed. New York: Eaton and Mains, 1904.

Bevan, Edwyn Robert. *Sibyls and Seers: A Survey of Some Ancient Theories of Revelation and Inspiration*. London: G. Allen & Unwin, 1928.

Bockmühl, Klaus. *The Unreal God of Modern Theology: Bultmann, Barth, and the Theology of Atheism: A Call to Recovering the Truth of God's Reality*. Colorado Springs, CO: Helmers & Howard, 1988.

Bohlmann, Ralph A. *Principles of Biblical Interpretation in the Lutheran Confessions*. St. Louis: Concordia Publishing House, 1968.

Bombaro, John. "Biblicism and the Imminent Death of American Evangelicalism." In *Built on the Foundation of the Apostles and the Prophets: Sola Scriptura in Context: The Second International Symposium on Lutheran Theology*. Edited by Tapani Simojoki. Evangelical Lutheran Church of England, 2013.

Braaten, Carl E. *History and Hermeneutics*. Philadelphia: Westminster Press, 1966.

———, ed. *Rightly Handling the Word of Truth: Scripture, Canon, and Creed*. Delphi, NY: ALPB Books, 2015.

Breck, John. "Exegesis and Interpretation: Orthodox Reflections on the 'Hermeneutic Problem'." *St. Vladimir's Theological Quarterly* 27:2 (1983): 75–95.

Brooks, Peter Newman, ed. *The Seven-headed Luther: Essays in Commemoration of a Quincentenary*. Oxford: Clarendon Press, 1983.

Brown, Raymond E. *The Critical Meaning of the Bible*. New York: Paulist Press, 1981.

Brunner, Emil. *Revelation and Reason: The Christian Doctrine of Faith and Knowledge.* Translated by Olive Trad Wyon. Philadelphia: Westminster Press, 1946.

Brunner, Peter. "Commitment to the Lutheran Confession—What Does It Mean Today?" *Springfielder* 33:3 (1969): 4–14.

Buchanan, James. *Modern Atheism: Under its Forms of Pantheism, Materialism, Secularism, Development, and Natural Laws.* Boston: Gould and Lincoln, 1857.

Bultmann, Rudolf. *Jesus Christ and Mythology.* Upper Saddle River, NJ: Prentice Hall, 1958.

———. *New Testament and Mythology and Other Basic Writings.* Translated by Schubert Miles Ogden. Philadelphia: Fortress Press, 1984.

Burtt, Edwin A. *The Metaphysical Foundations of Modern Physical Science: A Historical and Critical Essay.* 2nd rev. ed. London: Routledge and Kegan Paul, 1932.

Caldwell, Philip. *Liturgy as Revelation: Re-Sourcing a Theme in Twentieth-Century Catholic Theology.* Minneapolis: Fortress Press, 2014.

Calvin, John. *Institutes of the Christian Religion.* 2 vols. The Library of Christian Classics. Edited by John T. McNeill. Translated by Ford Lewis Battles. Philadelphia: Westminster Press, 1960.

"Cambridge University Church Society: The Inspiration of Holy Scripture." *Cambridge Review* 5 (1884): 376. https://books.google.com/books?id=FlFIAAAAYAAJ.

Chafer, Lewis Sperry. *Systematic Theology.* 8 vols. Dallas: Dallas Seminary Press, 1947.

Chemnitz, Martin. *Examination of the Council of Trent.* 4 vols. Translated by Fred Kramer. St. Louis: Concordia Publishing House, 1978.

———. *The Two Natures in Christ.* Translated by Jacob A. O. Preus. St. Louis: Concordia Publishing House, 1971.

Cohen, Norman J., ed. *The Fundamentalist Phenomenon: A View from Within; A Response from Without.* Grand Rapids: Eerdmans, 1990.

Comfort, Ray. *Luther Gold: Pure Refined.* Edited by Mary Ruth Murray. Alachua, FL: Bridge Logos Foundation, 2009.

Cragg, G. R. *Reason and Authority in the Eighteenth Century.* Cambridge: Cambridge University Press, 1964.

Craig, William Lane. " 'Men Moved by the Holy Spirit Spoke from God' (2 Peter 1.21): A Middle Knowledge Perspective on Biblical Inspiration." In *Oxford Readings in Philosophical Theology: Providence, Scripture, and Resurrection.* Vol. 2. Edited by Michael Rea. New York: Oxford University Press, 2009. http://www.leaderu.com/offices/billcraig/docs/inspiration.html.

Dragseth, Jennifer Hockenbery, ed. *The Devil's Whore: Reason and Philosophy in the Lutheran Tradition.* Philadelphia: Fortress Press, 2011.

Ebeling, Gerhard. *Word and Faith.* Translated by James W. Leitch. Philadelphia: Fortress Press, 1963.

Eggold, Henry J. "A Man's View of the Word of God Determines His Attitude toward Inerrancy." Faculty study paper. Copy in Concordia Theological Seminary Library, Fort Wayne, IN.

Engelder, Theodore. *Haec Dixit Dominus: Thus Saith the Lord.* St. Louis: Concordia Publishing House, 1947.

———. "Holy Scripture or Christ?" 2 Parts. *Concordia Theological Monthly* 10:7–8 (1939): 491–506.

———. *Reason or Revelation.* St. Louis: Concordia Publishing House, 1941.

———. "Verbal Inspiration: A Stumbling Block for the Jews and Foolishness to the Greeks." 16 parts. *Concordia Theological Monthly* 12:4–13:12 (1941–1942).

Eugene IV, Pope. "*Cantate Domino.*" Papal Bull from Council of Florence. http://catholicism.org/cantate-domino.html.

Fabro, Cornelio. *God in Exile: Modern Atheism: A Study of the Internal Dynamics of Modern Atheism, from its Roots in the Cartesian Cogito to the Present.* Translated by Arthur Gibson. Mahwah, NJ: Paulist Press, 1968.

Fackre, Gabriel. *The Doctrine of Revelation: A Narrative Interpretation.* Grand Rapids: Eerdmans, 1997.

Fairbairn, A. M. *The Place of Christ in Modern Theology.* London: Hodder and Stoughton, 1893.

Fairweather, A. M. *The Word as Truth: A Critical Examination of the Christian Doctrine of Revelation in the Writings of Thomas Aquinas and Karl Barth.* London: Lutterworth Press, 1944.

Farkasfalvy, Denis M. *Inspiration and Interpretation: A Theological Introduction to Sacred Scripture.* Washington, D.C.: Catholic University of America Press, 2010.

Flacius, Matthias. *How to Understand the Sacred Scriptures.* Translated by Wade R. Johnston. Saginaw, MI: Magdeburg Press, 2011.

Forbes, Alexander Penrose. *A Short Explanation of the Nicene Creed.* Oxford: John Henry Parker, 1852.

Fuerbringer, L. *Theological Hermeneutics: An Outline for the Classroom*. St. Louis: Concordia Publishing House, 1924.

Gaillardetz, Richard. *By what Authority?: A Primer on Scripture, the Magisterium, and the Sense of the Faithful*. Collegeville, MN: Liturgical Press, 2003.

Gaukroger, Stephen. *The Emergence of a Scientific Culture: Science and the Shaping of Modernity 1210–1685*. Oxford: Clarendon Press, 2006.

Geisler, Norman L., ed. *Biblical Errancy: An Analysis of its Philosophical Roots*. Eugene, OR: Wipf and Stock, 1981.

———, ed. *Inerrancy*. Grand Rapids: Zondervan, 1980.

Gensichen, Hans-Werner. *We Condemn: How Luther and 16th-Century Lutheranism Condemned False Doctrine*. Translated by Herbert J. A. Bouman. St. Louis: Concordia Publishing House, 1967.

Gerrish, B. A. *Grace and Reason: A Study in the Theology of Luther*. Oxford: Clarendon Press, 1962.

Gibbs, Jeffrey A. *Jerusalem and Parousia: Jesus' Eschatological Discourse in Matthew's Gospel*. St. Louis: Concordia Publishing House, 2000.

Gieschen, Charles A. "The Relevance of the *Homologoumena* and *Antilegomena* Distinction for the New Testament Canon Today: Revelation as a Test Case." *CTQ* 79:3–4 (2015): 279–300.

Gillespie, Michael Allen. *The Theological Origins of Modernity*. Chicago: University of Chicago Press, 2008.

Goldsworthy, Graeme. *Gospel-Centered Hermeneutics: Foundations and Principles of Evangelical Biblical Interpretation*. Downers Grove, IL: InterVarsity Press, 2006.

Grant, Edward. *God and Reason in the Middle Ages*. New York: Cambridge University Press, 2001.

Green, Lowell. "Toward a new Lutheran Dogmatics." *CTQ* 50:2 (1986): 109–117.

Gregory the Great. *Commentary on the Book of Blessed Job*. http://faculty.georgetown.edu/jod/texts/moralia1.html.

Grenz, Stanley J. and Roger E. Olson. *20th-Century Theology: God & the World in a Transitional Age*. Downers Grove, IL: InterVarsity Press, 2010.

Grubbs, Norris C. and Curtis Scott Drumm. "What Does Theology Have to Do with the Bible? A Call for the Expansion of the Doctrine of Inspiration." *Journal of Evangelical Theological Society* 53:1 (2010): 65–79.

Hagen, Kenneth. *Foundations of Theology in the Continental Reformation: Questions of Authority*. Milwaukee: Marquette University Press, 1974.

Hahn, Scott W. "For the Sake of our Salvation: The Truth and Humility of God's Word." *Letter & Spirit* 6 (2010): 21–45.

Haldane, Robert. *Authenticity and Inspiration of the Holy Scriptures Considered.* Edinburgh: John Lindsay, 1827.

Hamann, H. P. "A Plea for Commonsense in Exegesis." *CTQ* 24:2 (April 1978): 115–129.

Hannah, John D., ed. *Inerrancy and the Church.* Chicago: Moody Press, 1984.

Harris, James. *Analytic Philosophy of Religion.* Vol. 3 of *Handbook of Contemporary Philosophy of Religion.* Boston: Springer Science & Business Media, 2013.

Harrison, Peter. *The Bible, Protestantism, and the Rise of Natural Science.* Cambridge: Cambridge University Press, 2001.

Harrisville, Roy A. and Walter Sundberg. *The Bible in Modern Culture: Baruch Spinoza to Brevard Childs.* Grand Rapids: Eerdmans, 2002.

Harvey, Van A. *The Historian and the Believer: The Morality of Historical Knowledge and Christian Belief.* Philadelphia: Westminster Press, 1966.

Hausmann, William John. *Science and the Bible in Lutheran Theology: From Luther to the Missouri Synod.* Washington, D.C.: University Press of America, 1978.

Hegg, Tim. "The Role of Women in the Messianic Assembly." TorahResource, 1988. http://www.torahresource.com/EnglishArticles/RoleofWomen.pdf.

Henry, Carl F. H. *God, Revelation, and Authority: God Who Speaks and Shows: Fifteen Theses, Part Two.* Vol. 3 of 6. Waco, TX: Word Books, 1979.

———. "Narrative Theology: An Evangelical Appraisal." *Trinity Journal* 8:1 (1987): 3–19.

Herberger, Valerius. *The Great Works of God: Parts One and Two: The Mysteries of Christ in The Book of Genesis, Chapters 1–15.* Translated by Matthew Carver. St. Louis: Concordia Publishing House, 2010.

Hill, Christopher. *Antichrist in Seventeenth-Century England.* New York: Oxford University Press, 1971.

Hoenecke, Adolf. *Evangelical Lutheran Dogmatics.* 3 vols. Translated by James Langebartels and Heinrich Vogel. Milwaukee: Northwestern Publishing House, 2009.

Holmes, Michael W., ed. *The Apostolic Fathers: Greek Texts and English Translations.* Grand Rapids: Baker Books, 1999.

Horne, Thomas Hartwell. *An Introduction to the Critical Study and Knowledge of the Holy Scriptures.* Vol. 1. London: T. Cadell, 1828. https://books.google.com/books?id=kXVAAAAAcAAJ.

Hove, E. *Christian Doctrine.* Minneapolis: Augsburg, 1930.

Janse, Wim. "Reformed Antisocinianism in Northern Germany: Ludwig Crocius' *Antisocinismus Contractus* of 1639." *Perichoresis* 3:1 (2005). Republished, http://www.emanuel.ro/wp-content/uploads/2014/06/P-3.1-2005-Wim-Janse-Reformed-Antisocianism-in-Northern-Germany.pdf.

Jelf, William Edward. *Supremacy of Scripture: An Examination into the Principles and Statements Advanced in the Essay on the Education of the World.* London: Saunders, Otley, & Co., 1861. https://books.google.com/books?id=oWQXAAAAYAAJ.

Jenson, Robert W. *On the Inspiration of Scripture.* Delphi, NY: ALPB Books, 2012.

Jordahl, Leigh D. "The Theology of Franz Pieper: A Resource for Fundamentalistic Thought Modes among American Lutherans." *Lutheran Quarterly* 23 (1971): 118–137.

Kalomiros, Alxander. *Against False Union: Humble Thoughts of an Orthodox Christian Concerning the Attempts for Union of the One Holy, Catholic, and Apostolic Church with the So-called Churches of the West.* Translated by George Gabriel. 2nd rev. ed. Seattle: St. Nectarios Press, 1990.

Kantzer, Kenneth S. "Revelation and Inspiration in Neo-Orthodox Theology, Part II: The Method of Revelation." *Bibliotheca Sacra* 115.459 (1958): 120–127.

Kittel, Gerhard and Geoffrey W. Bromiley, eds. *Theological Dictionary of the New Testament.* Translated by Geoffrey W. Bromiley. Grand Rapids: Eerdmans, 1964.

Klein, Ralph. "How My Mind Has Changed." Conference: Building the One Foundation: Seminex at 35. 2009. http://www.lstc.edu/media/seminex/06-24-09-faculty-panel-Klein.pdf.

Kloha, Jeffery. "Kloha's Response to Montgomery Essay." *Christian News* 52:4 (Monday, Jan. 27, 2014): 13–15.

———. "Text and Authority: Theological and Hermeneutical Reflections on a Plastic Text." Listening to God's Word, Nov. 7–9, 2013 in Oberursel. Unpublished paper. http://steadfastlutherans.org/wp-content/uploads/2014/02/Text-and-Authority.pdf.

Klug, Eugene F. "Discussion Outline on the *Sola Scriptura* Essay." Faculty study paper. Copy in Concordia Theological Seminary Library, Fort Wayne, IN, 1968.

———. "Holy Scripture: The Inerrancy Question and Hermann Sasse." *Concordia Journal* 11:4 (1985): 124–127.

———. "Luther and Higher Criticism." *Springfielder* 38:3 (1974): 212–217.

———. "Review of *The End of the Historical-Critical Method.*" *Springfielder* 38:4 (1975): 289–302.

———. "Word and Scripture in Luther Studies Since World War II." *Trinity Journal* 5 (1984): 3–46.

Köberle, Adolf. *The Quest for Holiness: A Biblical, Historical and Systematic Investigation.* Translated by John C. Mattes. Minneapolis: Augsburg, 1936. Reprint, Eugene, OR: Wipf and Stock.

Kolb, Robert and Timothy Wengert, eds. *The Book of Concord: The Confessions of the Evangelical Lutheran Church.* Minneapolis: Augsburg Fortress, 2000.

Köstlin, Julius. *The Theology of Luther in its Historical Development and Inner Harmony.* 2 vols. Translated by Charles E. Hay. Philadelphia: Lutheran Publication Society, 1863.

Kramer, Fred. "The Inerrancy of Scripture: How it has been Understood, Attacked, and Defended." Faculty study paper. Copy in Concordia Theological Seminary Library, Fort Wayne, IN, 1964.

Krauth, Charles P. *The Conservative Reformation and Its Theology.* Philadelphia: General Council Publication Board, 1871. Reprint, Minneapolis: Augsburg, 1963.

Krentz, Edgar. *The Historical-Critical Method.* Guide to Biblical Scholarship: New Testament Guides. Edited by Dan O. Via, Jr. Philadelphia: Fortress Press, 1975.

Lee, Hoon. "Accommodation: Orthodox, Socinian, and Contemporary." *Westminster Theological Journal* 75 (2013): 335–348.

Leo XIII, Pope. "*Providentissimus Deus:* On the Doctrine of the Modernists." Encyclical, 1893. http://w2.vatican.va/content/leo-xiii/en/encyclicals/documents/hf_l-xiii_enc_18111893_providentissimus-deus.html.

Lewis, C. S. "Modern Theology and Biblical Criticism." *Brigham Young University Studies* 9 (1968): 35–48. Reprinted from *Christian Reflections,* edited by Walter Hooper. Grand Rapids: Eerdmans, 1967. https://ojs.lib.byu.edu/spc/index.php/BYUStudies/article/viewFile/4342/3992.

Lewis, Gordon Russell and Bruce A. Demarest, eds. *Challenges to Inerrancy: A Theological Response.* Chicago: Moody Press, 1984.

Library of Universal Knowledge: A Reprint of the Last Edinburgh and London Edition of Chambers's Encyclopædia. Vol. 6 of 15. New York: American Book Exchange, 1880.

Linnemann, Eta. *Historical Criticism of the Bible: Methodology or Ideology?* Grand Rapids: Baker Books, 1990.

Livingston, James C. *Modern Christian Thought, Vol. 1: The Enlightenment and the Nineteenth Century.* 2nd ed. Upper Saddle Creek, NJ: Prentice Hall, 1997.

Locke, John. *An Essay Concerning Humane Understanding.* Vol. 2. Edited by Steve Harris and David Widger. www.gutenberg.org/files/10616/10616.txt, 2004.

Long, Edward LeRoy Jr. *Religious Beliefs of American Scientists.* Philadelphia: Westminster Press, 1952. Reprint, Westport, CT: Greenwood Press, 1971.

Lotz, David W. "Luther and *Sola Scriptura.*" In *And Every Tongue Confess: Essays in Honor of Norman Nagel on the Occasion of His Sixty-fifth Birthday.* Edited by Gerald S. Krispin and Jon D. Vieker. Dearborn, MI: The Nagel Festschrift Committee, 1990.

Luther, Martin. *Bethlehem and Calvary: The Christmas and Easter Book of Dr. Martin Luther: A Thorough Exposition of Chapters 9 and 53 of the Book of the Prophet Isaiah.* Translated by Kenneth K. Miller. [K. K. Miller], 1988.

———. *Complete Sermons of Martin Luther.* 7 vols. Edited by John Nicholas Lenker and Eugene F. A. Klug. Grand Rapids: Baker Books, 2000. Vols. 1–4 published as *Sermons of Martin Luther: The Church Postils.* 8 vol. in 4 vols., 1995. Vols. 5–7 published as *Sermons of Martin Luther: The House Postils.* 3 vols. 1996.

———. *Dr. Martin Luthers Sämmtliche Schriften.* St. Louis: Concordia Publishing House, [1880–1910].

———. *Luther's Works.* Edited by Jaroslav Pelikan and Helmut Lehmann. 56 vols. St. Louis: Concordia Publishing House; Philadelphia: Fortress Press, 1955–86.

———. *The Table Talk of Martin Luther.* Translated by William Hazlitt. Philadelphia: The Lutheran Publication Society, 1873. http://cat.xula.edu/tpr/works/tabletalk/.

The Lutheran Church—Missouri Synod. "The Official Stylebook." Updated Jan. 2016. http://www.lcms.org/Document.fdoc?src=lcm&id=847.

McDonald, H. D. *Theories of Revelation: An Historical Study 1700–1960.* Grand Rapids: Baker Books, 1979.

McGrath, Alister E. *Intellectuals Don't Need God and Other Modern Myths: Building Bridges to Faith through Apologetics.* Grand Rapids: Zondervan, 1993.

McKim, Donald K., ed. *The Authoritative Word: Essays on The Nature of Scripture.* Eugene, OR: Wipf and Stock, 1983.

McLelland, Joseph C. *Prometheus Rebound: The Irony of Atheism.* Waterloo, ON: Wilfrid Laurier University Press, 1988.

Maier, Gerhard. *Biblical Hermeneutics.* Translated by Robert Yarbrough. Wheaton, IL: Crossway, 1994.

———. *The End of the Historical-Critical Method.* Translated by Edwin W. Leverenz and Rudolph F. Norden. St. Louis: Concordia Publishing House, 1977. Reprint, Eugene, OR: Wipf and Stock, 2001.

Marquart, Kurt E. *Anatomy of an Explosion: Missouri in Lutheran Perspective.* Fort Wayne, IN: Concordia Theological Seminary Press, 1977.

———. *Marquart's Works*. 10 vols. Edited by Herman J. Otten. New Haven, MO: Lutheran News, 2014–15.

Mattes, Mark C. "A Review of *Let Christ be Christ: Theology, Ethics & World Religions in the Two Kingdoms. Essays in Honor of the Sixty-Fifth Birthday of Charles L. Manske*." *Journal of Lutheran Ethics* 2:6 (2002). http://www.elca.org/JLE/Articles/954.

Maxfield, John, ed. *The Bible in the History of the Lutheran Church: The Pieper Lectures*. St. Louis: Concordia Historical Institute, 2005.

Miller, John. *The Divine Authority of Holy Scripture Asserted: From its Adaptation to the Real State of Human Nature, in Eight Sermons*. Oxford: W. Baxter, 1817. https://books.google.com/books?id=lllKAAAAYAAJ.

Moeller, Elmer. "The Meaning of Confessional Subscription." *Springfielder* 38:4 (1974): 193–211.

Montgomery, John Warwick. *Crisis in Lutheran Theology: The Validity and Relevance of Historic Lutheranism vs. Its Contemporary Rivals*. Vol. 1. Grand Rapids: Baker Book House, 1967.

Morgan, Michael L. *Dilemmas in Modern Jewish Thought: The Dialectics of Revelation and History*. Bloomington: Indiana University Press, 1992.

Muller, Richard A. *After Calvin: Studies in the Development of a Theological Tradition*. Oxford: Oxford University Press, 2003.

———. *Post-Reformation Reformed Dogmatics: Prolegomena to Theology*. Vol. 1. Grand Rapids: Baker Books, 1987.

Nafzger, Peter H. *These Are Written: Toward a Cruciform Theology of Scripture*. St. Louis: Wipf and Stock, 2013.

Naumann, Martin. "Messianic Mountaintops." *Springfielder* 39:2 (1975): 5–72.

Nichols, Stephen J. and Eric T. Brandt. *Ancient Word, Changing Worlds: The Doctrine of Scripture in a Modern Age*. Wheaton, IL: Crossway, 2009.

Noland, Martin R. "Walther and the Revival of Confessional Lutheranism." *CTQ* 75:3–4 (2011): 195–216.

Pieper, Francis. *Christian Dogmatics*. 4 vols. Translated by T. Engelder, J. T. Mueller and W. W. F. Albrecht. St. Louis: Concordia Publishing House, 1953.

———. *What is Christianity? And Other Essays*. Translated by John Theodore Mueller. St. Louis: Concordia Publishing House, 1933. Reprint, Malone, TX: Repristination Press, 1997.

Piepkorn, Arthur Carl. "The Significance of the Lutheran Symbols for Today: 1954 Faculty Lecture Series." In *The Sacred Scriptures and the Lutheran Confessions: Selected Writings of Arthur Carl Piepkorn*. Vol. 2. Edited by Philip J. Secker. Mansfield, CT: CEC Press, 2007.

Pinnock, Clark H. *A Defense of Biblical Infallibility*. Phillipsburg, NJ: Presbyterian and Reformed Publishing Company, 1967.

Piper, John. "A Reply to Gerhard Maier: A Review Article." *Journal of the Evangelical Theological Society* 22:1 (1979): 79–85.

Pius X, Pope. "*Pascendi Dominici Gregis:* On the Doctrine of the Modernists." Encyclical, 1907. http://www.papalencyclicals.net/Pius10/p10pasce.htm.

Porter, J. L. "Science and Revelation: Their Distinctive Provinces." In *Science and Revelation: A Series of Lectures in Reply to the Theories of Tyndall, Huxley, Darwin, Spencer, etc.* New York: Scribner, Welford & Armstrong, 1875.

Posset, Franz. "John Bugenhagen and the *Comma Johanneum*." *CTQ* 49:4 (1995): 245–251.

Preus, J. A. O. "The New Testament Canon in the Lutheran Dogmaticians." *Springfielder* 25:1 (1961): 8–33.

Preus, Robert D. *Inspiration of Scripture: A Study of the Theology of Seventeenth Century Lutheran Dogmaticians*. Mankato, MN: Lutheran Synod Book, 1955.

———. *The Theology of Post-Reformation Lutheranism: A Study of Theological Prolegomena*. 2 vols. St. Louis: Concordia Publishing House, 1970.

———. "Walther and the Scriptures." *Concordia Theological Monthly* 32:11 (1961): 669–691.

Radmacher, Earl D. and Robert D. Preus., eds. *Hermeneutics, Inerrancy, and the Bible*. Grand Rapids: Zondervan, 1984.

Ramm, Bernard. *Protestant Biblical Interpretation: A Textbook of Hermeneutics for Conservative Protestants*. 3rd rev. ed. Grand Rapids: Baker Books, 1999.

———. *Special Revelation and the Word of God*. Grand Rapids: Eerdmans, 1961.

———. *The Witness of the Spirit: An Essay on the Contemporary Relevance of the Internal Witness of the Holy Spirit*. Grand Rapids: Eerdmans, 1959.

Rehwaldt, Traugott H. "The Other Understanding of the Inspiration Texts." *Concordia Theological Monthly* 43:6 (1972): 355–367.

Reu, Johann Michael. *Luther and the Scriptures*. Columbus: Wartburg Press, 1944. Reprint, *Springfielder* 24:3 (1960): 7–111.

Reumann, John, ed. *Studies in Lutheran Hermeneutics*. Philadelphia: Fortress Press, 1979.

Richardson, Alan. *The Bible in the Age of Science*. London: SCM Press, 1961.

Rolston, Holmes, III. *Science and Religion: A Critical Survey*. Philadelphia: Temple University Press, 1987.

Rose, Matthew. "Karl Barth's Failure." *First Things* 244 (2014): 39–44.

Rose, Seraphim. *Genesis, Creation, and Early Man: The Orthodox Christian Vision*. Edited by H. Damascene. Platina, CA: St. Herman of Alaska Brotherhood. Rev. ed., 2011.

Rottmann, Erik. "Sermon of the Western Missouri Pastoral Conference: Preached at St. Paul High School, Concordia, MO" Unpublished, Oct. 17, 2006.

Ruokanen, Miikka. *Doctrina divinitus inspirata: Martin Luther's Position in the Ecumenical Problem of Biblical Inspiration*. Helsinki: Luther-Agricola Society, 1985.

Sagan, Carl. *The Demon-Haunted World: Science as a Candle in the Dark*. New York: Random House, 1995.

Sasse, Herman. Letter to Dr. J. A. O. Preus. Feb. 24, 1970. Copy in Concordia Theological Seminary Library, Fort Wayne, IN.

———. *Scripture and the Church: Selected Essays of Herman Sasse*. Edited by Jeffrey J. Kloha and Ronald R. Feuerhahn. St. Louis: Concordia Seminary, 1995.

Scaer, David P. *The Apostolic Scriptures*. St. Louis: Concordia Publishing House, 1971.

———. "Biblical Inspiration in Trinitarian Perspective." *Pro Ecclesia* 14:2 (2005): 143–160.

———. *An Introduction to the Method and Practice of Lutheran Theology*. Fort Wayne, IN: Concordia Theological Seminary Press, 1990.

———. *Ordination: Human Rite or Divine Ordinance*. Fort Wayne, IN: Concordia Theological Seminary Press. http://www.ctsfw.net/media/pdfs/ScaerOrdinationHumanRiteorDivineOrdinance.pdf.

———. "Reformed Exegesis and Lutheran Sacraments: Worlds in Conflict." *CTQ* 64:1 (2000): 3–20.

———. "The Theology of Robert David Preus and His Person: Making a Difference." *CTQ* 74:1 (2010): 75–91.

Scaer, Peter J. "The Gospel of Luke and the Christology of Martyrdom." Unpublished paper presented at the Fort Wayne Exegetical Symposium, 2003. http://static1.1.sqspcdn.com/static/f/38692/333371/1273663586093/The+Gospel+of+Luke.

———. "Jesus and the Woman at the Well: Where Mission Meets Worship." *CTQ* 67:1 (2003): 3–18.

———. *The Lukan Passion and the Praiseworthy Death*. Sheffield: Sheffield Phoenix Press, 2005.

Schaff, Philip. *Creeds of Christendom*. 3 vols. Rev. ed. Grand Rapids: Baker Books, 1984.

Scharf, Uwe Carsten. *The Paradoxical Breakthrough of Revelation: Interpreting the Divine-Human Interplay in Paul Tillich's Work, 1913–1964*. New York: Walter de Gruyter, 1999.

Schmid, Heinrich. *The Doctrinal Theology of the Evangelical Lutheran Church*. Translated by Charles A. Hay and Henry E. Jacobs. Minneapolis: Augsburg, 1875. Reprint, Philadelphia: United Lutheran Publishing, 1961.

Simpson, James Y. *Landmarks in the Struggle between Science and Religion*. London: Kennikat Press, 1925. Reprint, 1971.

Smith, Henry Preserved. *Inspiration and Inerrancy: A History and a Defense*. Cincinnati: Robert Clarke & Co, 1893.

Spitz, Lewis W. "Discord, Dialogue, and Concord: The Lutheran Reformation's Formula of Concord." *CTQ* 43:3 (1979): 183–195.

Stephenson, John R. "'Inerrancy'—The *homousias* of Our Time." *Logia* 3:4 (1993).

———, ed. *Hermann Sasse: A Man for Our Times?* St. Louis: Concordia Publishing House, 1998.

Strong, Augustus Hopkins. *Systematic Theology: A Compendium and Commonplace-Book Designed for the Use of Theological Students*. 2nd ed. New York: A. C. Armstrong and Son, 1889.

Sundberg, Walter. "Wilhelm Löhe on Pastoral Office and Liturgy." *Word & World* 24:2 (2004): 190–197.

Surburg, Raymond. "Implications of the Historico-Critical Method in Interpreting the Old Testament: Part 1." *Springfielder* 26:1 (1962): 6–25.

Thielicke, Helmut. *The Evangelical Faith: Prolegomena: The Relation of Theology to Modern Thought Forms*. Vol. 1 of 3. Translated and edited by Geoffrey W. Bromiley. Edinburgh: T&T Clark, 1978.

———. *How Modern Should Theology Be?* Translated by H. G. Anderson. Philadelphia: Fortress Press, 1969.

Thisleton, Anthony C. *The Two Horizons: New Testament Hermeneutics and Philosophical Description*. Grand Rapids: Eerdmans, 1992.

Thomassen, Einar. *Canon and Canonicity: The Formation and Use of Scripture*. Copenhagen: Museum Tusculanum Press, 2010.

Thompson, Mark. *A Sure Ground on Which to Stand: The Relation of Authority and Interpretive Method in Luther's Approach to Scripture*. Waynesboro, GA: Paternoster, 2005.

Treier, Daniel J. "The Superiority of Pre-critical exegesis?: *Sic et Non*." *Trinity Journal* 24:1 (2003): 77–103.

Trigg, Jonathan D. *Baptism in the Theology of Martin Luther*. Leiden: Brill, 2001.

Tulga, Chester E. *The Case against Modernism.* Chicago: Conservative Baptist Fellowship, 1949.

Valen-Sendstad, Olav. *The Word That Can Never Die: A Scriptural Critique of Theological Trends.* St. Louis: Concordia Publishing House, 1966.

Visser 't Hooft, W. A. *The Fatherhood of God in an Age of Emancipation.* Philadelphia Press: Westminster Press, 1982.

Voelz, James W. *What Does This Mean?: Principles of Biblical Interpretation in the Post-Modern World.* 2nd rev. ed. St. Louis: Concordia Publishing House, 1997.

Walther, C. F. W. "The Evangelical Lutheran Church, The True Visible Church of God on Earth." In *Walther and the Church.* Edited by Wm. Dallmann, W. H. T. Dau and Th. Engelder. St. Louis: Concordia Publishing House, 1938.

———. *The True Visible Church: An Essay for the Convention of the General Evangelical Lutheran Synod of Missouri, Ohio, and other States for its Sessions at St. Louis, MO, October 31, 1866.* Translated by John Theodore Mueller. St. Louis: Concordia Publishing House.

———. "Walther's Evening Lectures on Inspiration, 1885–1886." Translated by Thomas Manteufel. Presented to the Walther Round Table, 2005–2007. http://www.lutheranhistory.org/waltherrt/wrt-inspiration.htm.

———. *Walther on the Church: Selected Writings of C. F. W. Walther.* Translated by John M. Drickamer. St. Louis: Concordia Publishing House, 1981.

———. *Walther's Works: Church Fellowship.* St. Louis: Concordia Publishing House, 2015.

Weikart, Richard. "The Troubling Truth about Bonhoeffer's Theology." *Christian Research Journal* 35:6 (2012). Republished, http://www.equip.org/PDF/JAF5356.pdf.

Westfall, Richard S. *Science and Religion in Seventeenth-Century England.* New Haven, CT: Yale University Press, 1958.

Wingren, Gustaf. *Theology in Conflict: Nygren, Barth, Bultmann.* Translated by Eric Wahlstrom. Philadelphia: Muhlenberg Press, 1958.

Zahm, J. A. *Evolution and Dogma.* Chicago: D. H. McBride, 1896. Reprint, Hicksville, NY: Regina Press, 1975.

Zaun, Allan Andrew. "A Study of the Idea of the Verbal Inspiration of the Scriptures with Special Reference to the Reformers and Post-Reformation Thinkers of the Sixteenth and Seventeenth Centuries." PhD diss. University of Edinburgh, 1937. https://www.era.lib.ed.ac.uk/handle/1842/10291.

Zimmerman, Paul A. *A Seminary in Crisis: The Inside Story of the Preus Fact Finding Committee.* St. Louis: Concordia Publishing House, 2007.

www.ingramcontent.com/pod-product-compliance
Lightning Source LLC
Chambersburg PA
CBHW050527300426
44113CB00012B/1990